Here Is a
Human Being

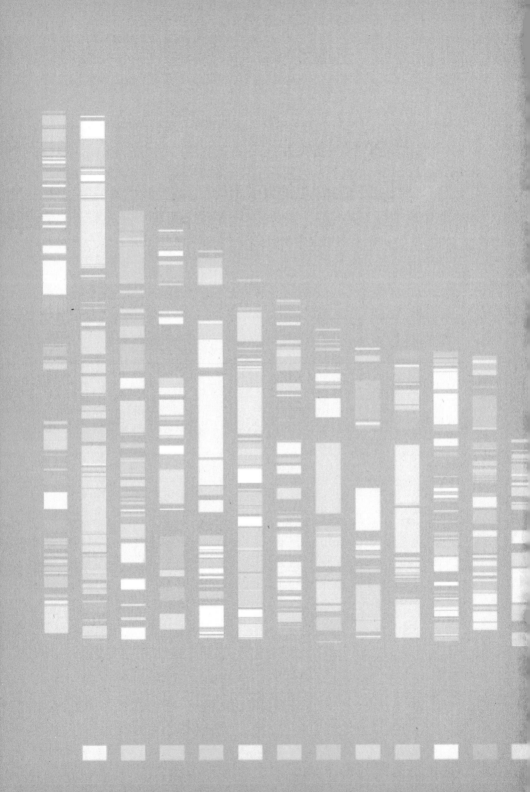

Here Is a Human Being

AT THE DAWN OF PERSONAL GENOMICS

MISHA ANGRIST

An Imprint of HarperCollins*Publishers*
www.harpercollins.com

HARPER

HarperCollins books may be purchased for educational, business, or sales promotional use. For information, please write: Special Markets Department, HarperCollins Publishers, 10 East 53rd Street, New York, NY 10022.

FIRST EDITION

Designed by Suet Yee Chong

Library of Congress Cataloging-in-Publication Data

Angrist, Misha.
   Here is a human being : at the dawn of personal genomics / Misha Angrist.
      p.   cm.
   Summary: "Misha Angrist, who has a Ph.D. in genetics and an MFA, brings us the first, inside story of the Personal Genome Project, its larger-than-life research subjects, as well as the political, social, and ethical issues that emerged throughout the study"—Provided by publisher.
   ISBN 978-0-06-162833-7 (hardback)
   1. Personal Genome Project. 2. Human genome—Databases. I. Title.
QH447.A54 2010
611'.0181663—dc22

2010009063

10 11 12 13 14   OV/RRD   10 9 8 7 6 5 4 3 2 1

For Ann

Three billion bases of DNA sequence can be put on a single compact disc and one will be able to pull a CD out of one's pocket and say, "Here is a human being; it's me!"

—WALTER GILBERT, 1992[1]

God could cause us considerable embarrassment by revealing all the secrets of nature to us: we should not know what to do for sheer apathy and boredom.

—JOHANN VON GOETHE[2]

When one goes on a journey of self-exploration, one should go heavily armed.

—PAUL VERLAINE[3]

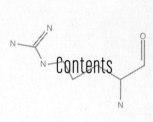

Contents

1

The Numerator

I was born in 1964. While I continue to make peace with social networking, I still think of myself as too old to be an organic part of the exhibitionism on YouTube, Facebook, Twitter, and MySpace. At best I am a poseur. I play in an occasional rock band with some other forty-somethings, and while we have a MySpace page,[1] none of us is terribly adept at using it. We have no idea, for example, how to program a "bot" to troll cyberspace and accumulate "friends" like so many chocolate Easter eggs; we have outsourced that kind of stuff to younger friends and to our children. We tend to be less mercenary than younger bands; we friend people we know and musicians we admire, never mind that in many cases they're dead (these days the definition of friendship doesn't even require that both parties be capable of drawing breath). In my band we will post pictures or movies of ourselves, but almost always with ambivalence, as we find them to be dorky, pretentious, or otherwise embarrassing. Like any collection of forty-five-year-olds still writing songs and performing them in dingy college-town bars on Friday nights for a dozen of their most patient, loyal, and sleepy friends, we are narcissists and stultifyingly vain, to be sure, but hey, at least we're discriminating.

As much as I want to hide my potbelly (probably a sign of insulin resistance and determined by genes acting in concert with ice cream), I agreed to make my genome—the DNA sequence that I inherited from my parents and that is uniquely mine just as yours is uniquely yours*—an open book. And if you want to study my phenotype—my health records, my ongoing struggles with depression and infatuation with selective serotonin reuptake inhibitors, or my dubious diet, for example—then go ahead, knock yourself out. It's all out there—with more to come—on the Internet. For doing this, I and the other participants in the Harvard-sponsored Personal Genome Project[2] have been lauded for our "bravery" by friends and colleagues, derided for our elitism and egotism by some social and genome scientists, and largely ignored by the medical establishment.

Are we—were we—really so brave? One of my favorite people who thinks about these things, Stanford law professor Hank Greely, said not long ago, "People believe in the magic of genes, and buy into the idea that they are the deepest secrets of our being. But maybe my credit card records come closer to being a deep secret of my being."[3]

As of mid-2010, few "whole" human genomes[†] had been sequenced—certainly no more than a few hundred. But in the next couple of years—or months—there will be thousands more, at least some of which will be widely shared with researchers or anyone who's interested. And the "genome-lite" version—having a half million or two million DNA markers analyzed for a few hundred bucks by private companies rather than a full sequence—is already a cheap commodity entrenched in the marketplace, much to the chagrin of many doctors and geneticists, and to the befuddlement of regulatory agencies.

---

* Unless you're an identical twin.

† Despite our best efforts, a "whole" human genome in 2010 is actually only about 93 percent; the other 7 percent has resisted our best efforts to sequence it.

What will this real-time experiment in science and radical openness mean? DNA sequencing may soon be cheap enough and reliable enough to make personal genomics as pervasive as cell phones, iPods, and LASIK surgery (assuming that, like those things, DNA sequences are actually useful).

But cheap sequencing and widespread sharing of genomic data will also bring with them "unintended consequences," card-carrying bioethicists' second favorite phrase behind "slippery slope." Nightmare scenarios include identity theft and loss of insurance.

On the other hand, even the leftiest and most passionate civil liberties advocates would have to concede that genetic discrimination is relatively rare: the two most notorious cases occurred several years ago. In one, Burlington Northern Santa Fe Railway was found to be surreptitiously testing its workers for a genetic variant thought to predispose to carpal tunnel syndrome (it doesn't—the company couldn't even figure out how to discriminate right); the railroad settled with thirty-six workers for $2.2 million in 2002.[4] In the other high-profile case, the Lawrence Berkeley National Laboratory was accused of having tested several of its employees for syphilis, sickle-cell anemia, and pregnancy without their consent. The lab settled in 1999, again for the magic sum of $2.2 million.[5] One presumes that in both cases the employers would have used positive test results (pregnancy?!) to justify denying the employees' insurance, raising their out-of-pocket costs for health benefits, or perhaps even firing them.

Again, two cases do not an epidemic make. But, as a former mentor of mine used to say, "Absence of evidence is not evidence of absence." Despite the lack of litigation, it may be that the temptation for employers to go all actuarial on their current and future employees is still too great. In 2005, for example, a higher-up at Walmart floated the idea of discouraging less healthy people from applying to work at the company, as a way of holding down health-care costs. In her memo, the vice president noted that Walmart workers tend to develop obesity-related—and partly genetic—diseases such as diabetes and heart disease at a higher rate than the national population (at the time, less than 45 percent of Walmart's employees had company health insurance).[6] More disturbing, in 2007 an exposé in the *Los Angeles Times* revealed that the U.S. military regularly practiced genetic discrimination. On a number of occasions servicemen and women with genetic conditions have found themselves kicked to the curb, with the armed forces arguing that in such cases they bear no responsibility for soldiers' health and disability benefits: you may have lost a limb in Fallujah, but your retinitis pigmentosa is preexisting, so, you know, *sorry*. Military doctors were said to discourage their patients from getting genetic tests.[7]

As I worked through drafts of this book, the world kept changing. The military has since revised its draconian policies on preexisting conditions.[8] And in May 2008, the Genetic Information Nondiscrimination Act (GINA) passed both houses of Congress and was signed by President George W. Bush.[9] Twelve years in the making, it was the end of a long and tortuous road for patient activists and a rare kumbaya moment in Washington in the post-9/11 Bush era.

But GINA was not and will not be a panacea: it is limited to employment and health insurance. It says nothing about life, disability, or long-term care insurance, all of which are likely to be of greater interest to insurance companies and to Alzheimer's patients and their families. Thus, if an insurer wants to deny you coverage or charge you exorbitant fees because you carry two copies of the APOE allele* that raises your risk of developing Alzheimer's fifteen-fold,[10] GINA can do nothing to influence those insurers.[11]

And its power over employers will be tested. Recently Pamela Fink of Fairfield, Connecticut, found out that she carried a mutation in BRCA2, one of the two most powerful hereditary breast cancer susceptibility genes. Despite years of glowing reviews from her employer, MXenergy,

---

* An allele is a version of a gene. For every gene, we inherit one allele from our mothers and one from our fathers; those alleles may or may not be the same. If they're the same, we are said to be *homozygous* for that allele. If they are different, we are said to be *heterozygous*.

after she shared her results with her bosses, she was demoted and eventually fired. [12]

And I was gonna put my entire DNA sequence on the Web?

In August 1996 I sat among a crowd of hundreds of other graduate students at Case Western Reserve University. As with most graduations, there was a palpable euphoria in the air. It was a rare cloudless summer day in Cleveland, an aging industrial city in northeast Ohio arguably most famous for a serpentine river that once caught fire. I made giddy small talk with the guy behind me in line as we waited our turn to make a brief procession across the stage, each of us feeling uncomfortable and a bit fraudulent in our caps and baggy gowns, anxious to shake hands with a dean we'd never met and take our elaborately adorned pieces of paper to be framed.

I wasn't going to win any prizes for originality or brilliance, but to my mother's and my own surprise, I had stuck it out and gotten my doctoral degree in genetics. I had spent the previous five years studying a rare birth defect called Hirschsprung's disease, named for the nineteenth-century Danish physician who first described it. It is also sometimes referred to by the even more unwieldy name "congenital aganglionic megacolon."[13]

Our gastrointestinal tract has its own primitive brain: the enteric nervous system.[14] Given that the adult intestinal tract is twenty-five feet long, it's not surprising that it requires some local neuronal signals to keep things moving along. Hirschsprung's disease occurs when, during early embryonic development, those neurons fail to take up residence in some part of the gut. Usually the few inches closest to the rectum are affected—this makes sense since those are the cells that will have traveled the farthest as the embryo grows.

The Hirschsprung's patient has great difficulty moving his bowels past the part without nerve cells—everything stalls. The bit just before the aganglionic segment swells, sometimes to the size of a football. Any competent pediatric surgery resident can look at an X-ray of a newborn

Hirschsprung baby's distended gut and make a reasonable guess at the diagnosis, which is confirmed by biopsy.[15]

Unlike most strongly genetic diseases, Hirschsprung's disease is highly treatable. Typically, once the diagnosis is made, a baby will have a colostomy procedure. He will then go home for six months to allow his intestines to grow. He will then return to the hospital, where the surgeon will perform a series of biopsies to locate exactly where in the gut the nerves stop. The surgeon will then resect the aganglionic part of the bowel. Six weeks later the surgeon will reverse the colostomy (some surgeons will dispense with the colostomy and try to repair the gut in a single surgery). Although most Hirschsprung's babies will grow up to lead normal lives, they are forever susceptible to severe constipation and potentially life-threatening infections of the bowel.[16]

Hirschsprung's is a stand-in for other complex diseases; that is, ones caused by some mixture of genes and the environment. I was part of one of two teams that found a major clue leading to the identification of the most important susceptibility gene.[17]

At some point during my first postdoctoral year, laboratory science began to lose its luster. I looked at those around me and saw the lives they were leading: the grant applications, the boring meetings, the parade of mind-numbing seminars, and the neurotic graduate students who required some mysterious combination of hand-holding, babysitting, thoughtful mentoring, and tough love. It was not nearly as much fun as it used to be.

And I was Hirschsprunged out. I took some measure of satisfaction in knowing that I was helping to elucidate the biology of the disease and that my work might play some small part in creating better diagnostics or therapies. But somehow it wasn't enough. As tragic diseases go, I suspect even most Hirschsprung's families would admit that their plight is not in the same league as those with a loved one suffering from Lou Gehrig's or late-stage lung cancer or Tay-Sachs. Instead, here was a genetic disease that was actually treatable or, as personal genomics partisans like to say, "actionable." At the time I wondered if anyone really cared about what we were doing. Genetic research has been—and continues to be—criticized for throwing lots of resources at rare diseases, and I suppose I worried that I was part of the problem. The genetic aspects of the disease I studied were fascinating, but we were a small, insular community studying a small, mostly treatable disease affecting one in five thousand kids.[18]

Meanwhile, molecular genetics was changing at a pace that, more and more, left me dizzy and exhausted. The seminars I attended may as well have been in Esperanto. I was falling behind. Science had revealed to me with much clarity that I was now a full-grown, thirty-three-year-old dinosaur. Staying in the game would mean assimilating massive amounts of biochemistry, deep and broad computational skills, or both. Genetics was becoming genomics, a digital science, and one ought to have had more of a quantitative clue than I did. I had a choice: I could change or die.

I decided to die.

I left laboratory genetics. I worked as a market researcher, feckless financial manager (I know those words seem redundant these days), biotech consultant, and science editor. Over the next seven years I rarely thought about Hirschsprung's disease. Once in a while I might stumble upon a paper by my advisor's group or other people I knew describing some novel finding that, if I really followed the paper trail, I might be able to trace back to my own work ten to fifteen years earlier.

But in 2005, Hirschsprung's reappeared in my life in a sudden, unlikely, and disturbing way, like a drunken ex turning up at one's wedding. Wielding a knife. On December 14, the eve of my mother's birthday, my nephew Jesse was born with a dilated colon. Three days later he was diagnosed with Hirschsprung's disease.

What did it mean? my family wondered aloud. I knew the question was both clinical and existential, though I could supply neither type of answer. What was going to happen to my brother's new son and my parents' last grandchild? I hadn't a clue. I immediately became the Expert, though I remembered little. I was as flummoxed as my family was, but out of love, hubris, and a desire to be a hero, I did not betray my ignorance. I took my thesis off the shelf for the first time in years. I sent emails and made phone calls to people who still "did" Hirschsprung's. I reread articles about pediatric surgery techniques, the details of which I'd once known cold.

As my heartbroken and sleep-deprived brother and sister-in-law watched their new baby go in and out of the hospital again and again, I was overtaken by a sort of numbness, a paralysis. How strange—how impossible!—that the rare disease I'd devoted eight years to understanding was suddenly no longer an intellectual abstraction, a genetic problem to be solved in a lab or on a computer. Someone in *my* family was to have five surgeries over the first year of his life, with still more to come. Someone— or someones—in *my* family had to change colostomy bags and purchase boatloads of strange dietary supplements and be on constant guard for signs of infection. I thought of my previous incarnation and the Hirschsprung families who would occasionally call the lab. I remembered their need to tell me their stories. I remembered feeling awkward at not being able to empathize. My two-plus years of training as a genetic counselor had failed me. I knew more about Bayesian risk estimates than I did about what to actually *say* to the grief-stricken mother of a sick newborn. I was mostly useless. But how could I possibly empathize? I was not one of them.

Now I wondered about Jesse. What if he carried one of the mutations I had identified ten years earlier?[19] What if my brother did? What if I did? And what *did* it mean?

God, I decided, was fucking with us.

I called my brother on the phone one day. He was tired and loopy after another night on the foldout chair in the hospital. "How's it going?" I said. An absurd question and a pretty feeble conversation starter.

"Why," my brother wanted to know, "couldn't you have done your graduate work on the gene for large penises?"

I tell the Hirschsprung's story not to elicit sympathy, though my brother and his wife surely deserve it in spades, as do all the parents who've ever had to watch their babies tethered to tubes in the PICU, terrified and uncertain about what the coming days might bring. Nor do I tell it because it strikes me as so metaphysically improbable, though, despite knowing better, I remain convinced that it is. I tell it because it points up the fact that all human genomics is personal—that is to say, it is finally about *us*. Mothers and fathers can negotiate almost anything: marriage, money, careers, sex, cooking, laundry, the Netflix queue, who gives the dog a bath. What they can't negotiate are their own genomes (although with techniques such as preimplantation genetic diagnosis, a few are beginning to negotiate the genomes of their children). Occasional strange cases notwithstanding, every parent gives his or her biological child 50 percent of that child's DNA. And every one of us, regardless of zip code, membership in an executive health program, or religious affiliation, carries at least a handful of harmful mutations that may or may not manifest in us or in our children, should they inherit them as part of that 50 percent. Yes, Jesse's Hirschsprung's disease was an unlikely event—on the order of one in five thousand. But unless we get hit by a bus or succumb to an infectious disease, eventually almost all of us are the numerator, the "one in" something—cancer, heart disease, diabetes, Alzheimer's. Genes are rarely the final arbiter of these late-onset chronic conditions, our understanding of them remains woefully inadequate, and they probably don't constitute the secrets of our being, but there's no denying the importance heredity plays in them.

One of the promises of personal genomics is that it will tell us exactly what we are at risk of becoming the numerator for. One of the dangers is that it might also tell our insurers the same thing while not being actionable. It will provoke suspicions and perhaps ulcers and force us to think about our destinies in terms of probabilities, as though we are watching the tote board at a Las Vegas sports book a few minutes before kickoff. *How will I die?* It might also tell us something about various "positive" traits—intelligence, memory, musical aptitude, athletic ability—and how we measure up . . . or down. What will we do when our entire genomes are no longer abstractions, but palpable bits of information we carry in our pockets?

A growing number of people are opting to find out. With some trepidation, I became one of them.

2

"His Own Drum"

Like a character on *Lost*, John Halamka has arrived from the future.

Wearing all black and brandishing his BlackBerry, he looks the part. And he lives it. As chief information officer at Harvard Medical School and CareGroup Healthcare System, he is responsible for the information needs of three thousand doctors and the health records of two million patients. He is conversant in twelve computer languages. He is a self-taught mycologist: if you find yourself in the ER after having ingested a wild mushroom, it is likely to be Halamka who gets called to figure out which of the twenty-five hundred North American mushroom species you ate and what to do about it.[1]

He is also apt to be mistaken for a shoplifter at Home Depot.

After the Food and Drug Administration approved the technology in 2004, Halamka had a bean-sized Radio Frequency Identification chip implanted in his right triceps. The RFID contains his medical record identifiers. In part his interest was personal: An avid mountain and ice climber, Halamka was concerned that if he fell off a cliff and lost consciousness, his would-be rescuers would not be able to identify him. Now by scanning his arm they could. The downside is that the same ISO-standard 134.2 KHz RFID scanner at the mall may occasionally confuse him with a roll of duct tape or a garden hose.[2]

Visit Halamka in his feng shui corner office on Harvard Medical School's Longwood campus and chances are you'll find him serenely clacking away at his computer, his desktop as clean as the day it left the office-furniture showroom. His lunch is all vegan, probably eaten with chopsticks and great dexterity. Send him an email and you'll likely get a response within minutes, just like the hundreds of others who've emailed him that day, including the ER docs wondering whether the mushroom little Billy swallowed is going to be lethal, hallucinogenic, or merely vomit-inducing. Listen to Halamka talk in his calm, measured cadences and it's not hard to imagine his flat Iowa accent booming out from your local public radio station.

When I visited, his office inside the Center for Educational Technology was emblematic of the rest of the medical school these days: an incongruous mix of Boston Brahmin stateliness and sleek hi-tech utility. Outside his door, banks of computers lined long tables in an otherwise open area, like a bullpen in a brokerage firm, only without the manic buying and selling, just the basal hum of conversation and the HVAC. Clad in his usual onyx ensemble ("At 3 AM halfway around the world, you do not have to think about color matching"[3]), hands folded on his lap, Halamka emphasized that his participation in the RFID trial was about more than just preparing for some hypothetical mountaineering accident. He saw it as both part of his job description and a moral obligation. Having the chip implanted allowed him to answer questions that might someday be relevant to his patients: Does an RFID hurt? (No, but there is some discomfort upon implantation.) Would it lead to infection? (Not so far.) Would it place any limitations on his activities? (Other than getting rung up for $27.99 at big-box retailers, no.)

There was the stalker, however. "This guy thought that RFID tagging was the mark of the beast. He basically devoted his life to blogging about me." Rather than antagonize his "fan," Halamka began a dialogue with him. "He said his name was Bob, he lived in Kansas City, he was disabled, and that following me was going to be his daily activity." After a while Bob went away.[4] Even the mark of the beast can get tedious.

Did the stalking episode dampen Halamka's appetite for early adoption? Hardly. Stalking him is now an utterly trivial exercise on par with checking the weather forecast: He has an application on his BlackBerry that tracks him via GPS and produces Google maps of his whereabouts. Anyone with a username and password can locate him anywhere on the planet.

"I think somebody has to do these experiments," he said of the Personal Genome Project. "We need to get a dataset and figure out what all the implications of this stuff could be."[5]

Halamka was probably destined to become Subject Number Two: like him, the PGP was another Harvard-based bit of self-experimentation and potential compromises in personal privacy. That's not to equate the two. For all of their potential applications, in the early 2000s RFIDs and GPS devices were really nothing more than LoJack for human beings—fancy dog tags. But if Harvard geneticist George Church had his way, the Personal Genome Project would raise the stakes by making public thousands of identifiable human DNA sequences. Church came to Halamka's attention several years ago after Church posted his own medical records online.[6]

> We would like to invite you to participate in a research study. You have been asked to participate because you are a healthy individual with sufficient training in human genetics and human subjects research to be able to give informed consent for a public and open-ended study.
>
> The main scientific goal of this study is to explore ways to connect human genotype and phenotype information,

i.e., human genome sequence, medical records, and non-medical physical traits. . . . The ethical and human goals include educating participants and the general public about the risks and potential alternative pathways that genetics can take. . . . We also hope to discover what consumers, clinicians, and researchers might want and not want and why.[7]

Researchers have long been interested in the relationship between genome and—in the trendy parlance—*phenome*, that is, how what's encoded in our DNA interacts with the environment to give rise to traits ranging from height to athletic prowess to behavior to assorted health conditions, be they cancers or various forms of male-pattern baldness.[8] Datasets have gotten larger and the DNA sequencing industry has become a multibillion-dollar behemoth.[9]

The Personal Genome Project's initial goal was to recapitulate the Human Genome Project; the HGP was the effort by both government and private-sector scientists to decode the entire complement of human DNA. The PGP would sequence the DNA of ten individuals, that is, ten persons' entire collection of 6 billion base pairs,* a task that by 2009 would take any properly equipped and extremely well-funded molecular genetics lab no more than a few weeks and a few hundred thousand dollars.[10] But statistically speaking, ten people is not that much different from one person or even zero people, which is to say, it's not much at all. Most if not all of the insights from those ten genomes would have to be replicated in a much larger sample, an undertaking that would not commence in earnest for a while yet. But in the early days when George Church discussed the PGP with me, he often began sentences with the phrase "When we get to a million."[11] Even though many of us rolled our eyes at that (and even its less ambitious interim successor, "When we get

---

* Six billion base pairs = three billion paired nucleotides—the A's, G's, T's, and C's of the DNA alphabet—on one double helix times two sets of chromosomes, one from Mom and one from Dad.

to a hundred thousand"), the premise was fundamentally correct: the real promise of genomics resides in large numbers. Only by studying thousands of people (and corralling a collection of supercharged computers and really smart people) can we begin to detect the subtle and meaningful DNA variants that affect complex and common traits like heart disease, arthritis, and diabetes. Those traits have been among us for millennia, but the technology capable of identifying their molecular underpinnings did not reach fruition until twenty-odd years ago. (Whether what we find will be at all helpful is another question entirely.)

Church, the PGP's Grand Pooh-Bah, is a tall man in his early fifties often hunched over the tiny netbook he carries with him everywhere. With his thick beard and backswept hair he reminds one of a healthier, pre-Appomattox Robert E. Lee crossed with a younger incarnation of the Band's keyboardist Garth Hudson, and perhaps a touch of Gandalf: a gnostic, gentle giant. He smiles and laughs easily. He is at peace with what he admits to be an idiosyncratic view of the world. When students and postdocs ask if they should share some juicy piece of data with other labs, his answer is almost invariably yes. His voice is a mellifluous baritone with a trace of twang; he has the air of a southern gentleman who's amused and beguiled by most of what he sees. His bright green eyes do not betray madness exactly, but the mischievous twinkle is undeniable.

According to his official online PGP profile, he's six feet four, 245 pounds. He takes statins (hyperlipidemia; his personal Web page cites a heart attack in 1994). He is narcoleptic. He has had squamous cell carcinoma. He takes vitamins.[12]

Every morning George walks the 0.8 miles from his house in Brookline to his office and lab in the unimaginatively named New Research Building in the Longwood Medical Area, a sleek glass edifice that could work as a setting for a science fiction movie. It is clean and new. Security is tight: two guards sit at the entrance. They greet visitors and monitor the goings-on via sixteen digital cameras. The restrooms are "smart": enter and the lights turn on. Upstairs they are identified by the male and female karyotypes: XY and XX chromosomes, respectively. George's office is spacious but not ostentatious. It's moderately bright: surrounded by glass on two sides, but nestled among tall buildings and parking garages. Children's Hospital is visible at the end of the access street two blocks away. Immediately below the window is a small green space that was recently invaded by a phalanx of geese ("More geese poop!" said George[13]). A welder's sparks tumbled from above—yet another parking garage was going up next door.

The culture of the Church lab is reminiscent of the best labs I worked in. It is characterized by an utter lack of pretentiousness. People wear jeans, T-shirts, and sandals or flip-flops along with their lab coats, goggles, and latex gloves. Elaborate drawings of molecular biology experiments appear on wipe boards next to cartoons of superheroes and deep-sea divers, and cutouts of rock stars. There is conversation, but mostly it is quiet. Everyone has his or her experiments to do. The British-born geneticist and Nobel laureate Oliver Smithies,[14] famous for discovering a bunch of things, including how to knock out single genes in mice in order to see what traits those genes control, continued—and for all I know, continues—to go to his lab at the University of North Carolina every day well into his eighties. When I asked him why, he likened the lab to a monastery and told me that good scientists more or less take a vow of poverty in order to do science: they must renounce most of the rest of their lives.[15] My sense is that George would never regard it as a solemn vow. Science is fun, so why not do it?

When George was born, his biological father, Henry Stewart McDonald III, was working as a clown. At various times he was a race car driver, actor, model, auto mechanic, aircraft pilot, movie producer, writer, TV commentator, and champion water-skier.[16] After a public address announcer introduced him as "Barefoot McDonald" because of his disdain for wearing shoes, the name stuck. The resistance to footwear, however, was not an act: A 1971 *Sports Illustrated* profile described his infamous ejection from a Las Vegas casino under protest. "Whaddaya mean, bare *feet*?

You've got broads in here wearing dresses without any backs. Some of the dresses don't even have any fronts."[17] In 1992, at the age of sixty-seven, McDonald was inducted into the Water Ski Hall of Fame.[18] Indeed, there is an adorable picture of seven-month-old George sitting on a tiny chair that has been fastened to a tabletop and is being pulled through the water by his father. "My mother was beside herself with terror," George recalled. "My father was convinced I was enjoying it."[19]

When George was two, McDonald left his mother, Virginia Anne Strong, who went on to marry twice more. George's wife Ting Wu attributed much of her husband's success to his mother's influence. "An amazing woman," she said of Strong. "She was brilliant—a lawyer, a psychologist, and an artist. And like George she was extremely flexible and optimistic. I don't think I ever heard her say a negative thing. I think George grew up knowing he could weather anything."[20]

George lived in Tampa and Clearwater, Florida, until he was a teenager. His second stepfather, a doctor and Phillips Academy Andover alum, suggested George might like the prestigious boarding school. George excelled at Andover, both in academics and on the varsity track and wrestling teams. He applied to and was accepted at Harvard and Duke, went to Duke, and finished an undergraduate degree in zoology and chemistry in two years. He received a predoctoral fellowship from the National Science Foundation and dove into graduate work in microbiology.

Within a year he had flunked out.

To hear George tell it, microbiology was the wrong home for someone with a passion for biochemistry. And when he switched to biochemistry, he found that his new department was not terribly interested in him, either: the orphan narrative, it seemed, had followed him to the academy. And then there was the course work. George often read his assigned textbooks cover to cover by the first or second week of class. After that, there wasn't much point in showing up; after all, he could be using that time to do actual hands-on research rather than simply read abstract accounts of what had already been done. He refused to take "baby science courses" just to fulfill curricular requirements. Recalling this, he gave a sheepish laugh and shrugged. "I guess I should've told them I wasn't going to attend classes."[21]

His academic problems did not prevent him from beginning to get his name on papers, including a first-author publication in *Proceedings of the National Academy of Sciences* at the age of twenty-two.[22] After Duke wished him luck elsewhere, he reapplied to Harvard, stated in his application he wanted to work with renowned molecular biologist Walter Gilbert on a new approach to decoding DNA, and was accepted. In an interview, Gilbert maintained that Harvard's admissions policy at the time did not consider incoming students' preferences for faculty, let alone require them. He was willing to believe, however, that George flunked out of Duke. "He always marched to his own drum."[23] George began his doctoral studies at Harvard in the fall of 1977, just a few months after Gilbert and graduate student Allan Maxam published a paper on how to sequence DNA,[24] for which Gilbert would go on to share in the 1980 Nobel Prize.[25]

It was at Harvard that George met his future wife and Harvard colleague Ting Wu, also then a graduate student in genetics. The two were in a class together on the structure of chromatin, the essential but still somewhat mysterious molecular scaffolding found in chromosomes, an irony not lost on her. "Chromatin is one of the aspects of inheritance not entirely coded for by DNA."[26]

Her first impressions? "I remember noticing that this guy knew everything. Someone mentioned that an interesting article had just come out but couldn't remember what journal or who the authors were. George always knew: 'It's in *PNAS*, volume such and such, page so and so.' But when he said those things it would never come off in a know-it-all kind of way. He was just a nice guy. And even when I first I met him he was fascinated by sequencing."[27]

"When I got to Wally's lab in '77 I was already thinking about it," George said. "It just wasn't rational thought."[28]

The first quasi-complete sequence assembled by the public Human Genome Project (HGP) in 2003* was actually a composite of several

---

* A draft version was first unveiled in 2000.

anonymous people who were recruited via ads in the *Buffalo News* in 1997.[29] One sample known as "RP11" appears to have been overrepresented (George told me that reporters have since gone to Buffalo to try to learn the identity of RP11[30]). In parallel to the publicly funded HGP, Craig Venter used his own DNA as part of a private sequencing effort led by the company he once ran, Celera,[31] and subsequently by his own research institute. Venter's entire sequence was published in 2007[32] and represented the first complete human reference genome from a single individual.* In any case, depending upon when one marks the start date, obtaining that first composite sequence took something like thirteen years and $3 billion in public funds plus the hundreds of millions spent by Celera and its stockholders.

Until 2006 or so, few gave the idea of personal genomics much consideration, rational or otherwise. Yes, sequencing had already gotten much cheaper by the early 2000s, but until about 2005 a human genome was still going to cost you tens of millions of dollars.

Since his arrival in Harvard Yard in the late 1970s, Church has spent his time there (and at various companies he's helped found) looking for ways to make sequencing cheaper and easier. In 2003, *Genome Technology* magazine suggested he would make "a good candidate for a lifetime achievement award in genome sequencing."[33] He and later his small army of students and postdocs found clever chemical shortcuts and cost-savers. Over the last few years, much of his lab's attention has been on polymerase colonies, or "polonies," a method that uses enzymes to amplify billions of short DNA fragments and stitch those together into a form that

---

* Jim Watson received his genome sequence on a hard drive in an elaborate ceremony at Baylor College of Medicine on May 31, 2007, while Venter's was published in the journal *PLoS Biology* in September 2007. But Venter, Watson's once and future rival since the early days of the Human Genome Project, made sure to deposit his own sequence in GenBank, the public DNA database, before Watson deposited his. At the Baylor event, Watson recalled the race to sequence the human genome and said of Venter, "He was clearly so bad we couldn't let him win." A paper describing Watson's sequence was published in *Nature* in 2008.

can be sequenced.[34] Polony technology has since been licensed to several companies. As we'll see, Church has teamed up with an engineering firm to make and sell a polony sequencer ("the Polonator") for significantly less than other companies in the sequencing business.[35] Among DNA sequencers, it was and is a thing of beauty: even the beta testers who couldn't get it to work admired its aesthetic qualities. The Polonator has a robotic arm inside humming along and dispensing chemicals, and a platform that holds the slide upon which sits the DNA. An expensive camera takes pictures of the samples. The whole apparatus sits inside a slightly Daliesque warped blue cube about the size of a small clothes dryer.

For the last few years sequencing costs have been in free fall, thanks in no small part to George's work. In 2007 he told me he expected that soon his lab would be able to use the Polonator to sequence the entire human exome, that is, the protein-coding 1 percent of the human genome, for as little as a thousand dollars. In reality this was still more than two years away, but I came to learn that this type of error is a chronic occupational hazard for an incurable optimist such as George.

The implications of an affordable exome and/or genome left him both excited and unsettled. And he wasn't the only one. He forged ahead with the PGP, but was forced to use private money. He believed the National Institutes of Health (NIH) was balking at paying for the project (despite having approved it and every other aspect of his $10 million genome technology grant) because he refused to do it under the standard ethical paradigm set forth by the agency, that is, one in which subjects give informed consent and in return are promised, more or less, privacy and confidentiality.[36] "I just feel uncomfortable signing up people under the supposition of anonymity if that's not something that can be assured."[37]

Why couldn't it be assured? On his website, George offered a laundry list of real-life scenarios where presumptively anonymous subjects were reidentified without their consent.[38] Some of these became infamous stories. In the 1990s, for example, then–MIT grad student Latanya Sweeney used publicly available voter records and a public, anonymized database of state employees to identify the medical records of Massachusetts governor William Weld. She was also able to identify the five African-American women living in the predominantly gay enclave of Provincetown, Massachusetts, purely on the basis of public data.[39] Another example: a few years ago, a fifteen-year-old boy used a combination of a commercial DNA test of his Y chromosome, genealogical records, and Internet searches to locate his "anonymous" sperm donor father.[40]

In George's eyes, these types of privacy-hacker stories would only become more common as genomic data proliferated. DNA is the ultimate digital identifier, after all—a social security number is only nine digits, while a genome is three billion. Not that anywhere near that number would even be necessary to hack one's identity: a paper in *Science* suggested that as few as eight genetic markers could be considered a risk for reidentifying humans.[41] Twenty-five such markers would likely be fully identifying, akin to the thirteen forensic DNA markers typically typed from crime scene samples.* And in 2008, researchers from the Translational Genomics Research Institute and University of California at Los Angeles were able to identify individual DNA samples from a complex mixture of samples from as many as two hundred people, even if a particular individual's sample accounted for only one-tenth of 1 percent of the total DNA mixture.[42] In response the NIH immediately backpedaled on its promise to facilitate widespread sharing of human DNA samples among the studies it funded.[43] George saw this coming. Given the technical realities, he said, to promise privacy and confidentiality would be disingenuous at best.[44]

And there's another reason not to do it, he said: it makes for bad, or at least limited, science. Ensuring anonymity, assuming it can be done at all, means restricting the use of any phenotypic information that could be used to reidentify a subject. But it is exactly those sorts of unique bits of data on human traits—hair color, eye color, facial features, cognitive

---

★ Forensic DNA markers as used by the FBI tend to be highly variable ("polymorphic"); that is, they are present in many different versions in the human genome and therefore carry more information than standard markers used in genetic studies, which typically come in just one of two "flavors."

measures—that George saw as necessary to fully leverage whole genomes and really begin to understand the human gestalt. "Some people look at a person's face and think they can tell everything about his past, present and future," he said. "That's not true, of course. But when you ask who you are, a huge fraction is what your face is. So take the face plus the genome, plus metabolites, plus proteomics, and then you're starting to get something that is without parallel in current practice."[45] Giving broad access can have practical benefits, too. One of George's favorite stories— recounted on his website, naturally[46]—is from 2004. He was preparing to give a lecture in Seattle at the University of Washington Medical School when a hematologist in the audience raised his hand and said, "You really ought to get your cholesterol checked." He had looked at George's personal medical records Web page and seen that his total cholesterol was 288 mg/dL (the normal level is less than 200). The 288 measurement was more than a year old and George had not been back to his doctor to see if the statin he was taking was having the desired effect. It turned out it was not. His doctor doubled the dose and George went back to a strict vegan diet. In six weeks his cholesterol had fallen to 156 mg/dL. What struck George was that this interaction with a total stranger had had a tangible positive effect on his health. "That total-stranger expert will eventually be replaced by software," he predicted.[47]

For that to happen and for it to be of practical value will require both genetic data and health records. George's Harvard colleague, pediatrician, and champion of electronic health records Isaac "Zak" Kohane told me essentially the same thing as every other genome scientist I spoke to. "Without the phenotype, the genome is just not that useful."[48] But even ambitious genome sequencing efforts such as the internationally sponsored "1000 Genomes Project" were not prepared to collect detailed trait data on their subjects, preferring instead to use "old" DNA samples from "anonymized" subjects who had consented years ago.[49] Collecting trait data is hard, it's time-consuming, it's expensive, and yes, it makes the subjects that much easier to identify.

For the last problem, George's solution was "open consent." Why not, he asked, recruit subjects willing to forgo guarantees of privacy and confidentiality—that is to say, people like him? This is why he became Subject Number One in the PGP.[50]

Of the first two prominent public genomes, Craig Venter has been open about his sequence, even though it told him that he is at increased risk for Alzheimer's and macular degeneration.[51] James Watson's DNA now resides in a public DNA database.[52] Watson's son Rufus has schizophrenia.[53] When I asked Watson if he consulted with his family members before making his genome public, he shook his head and smiled. "They might have said no."[54] Indeed, one sometimes wonders what might ultimately be found in Watson's DNA. The same week in which he publicly implied that Africans were less intelligent than whites and lost his job because of it,[55] he suggested that crystallographer and onetime rival the late Rosalind Franklin had been partially autistic.[56] One wants to kick him under the table or pull him aside and say, "Dude. Stop." When I asked George about Watson being one of the first complete and public genomes, he gave a less than ringing endorsement. "I don't think it was an ideal choice. If you're gonna put all your eggs in one basket, he doesn't seem like the obvious first basket."[57]

In 2005, the Harvard Medical School Institutional Review Board approved the Personal Genome Project, but only after lengthy discussion.* Initially, according to rabbi and IRB member Terry Bard, a longtime lecturer on pastoral counseling in psychiatry at Harvard, the Harvard IRB was not even sure the PGP was in its bailiwick. "Everyone, including me," said Bard, "was scratching their heads and saying, 'Why is this here?' A number of members were not convinced—and maybe remain unconvinced—that, as presented, this was actually a *research* study. Did it meet the basic scientific criteria for research?" George appeared before the IRB twice to

---

* The Institutional Review Board is the ethical review board charged with approving, monitoring, and reviewing biomedical research on humans. Every hospital and academic medical center has one or more IRBs; I serve on Duke's.

convince members that it was indeed research and to seek IRB guidance as to what to include in the protocol and the consent form.[58]

Initially, George was asked to limit the initial subject total to one: himself. Bard shepherded him through the consent process. Since George was putting both his genotype and phenotype on the Web and therefore making it available for public scrutiny, Bard said the IRB wanted to gauge the initial experience first. "We wanted to know how it would play in Peoria, so to speak," said Bard. "Dr. Church made monthly reports back to us on how his decision was being received and what kinds of interactions he had had."[59]

Bard is a short man who, when I met with him in his softly lit office at Beth Israel Deaconess Medical Center, wore a white hospital coat. As a perpetually confused Jew, I was anxious to avail myself of his other skill set. So at the end of our hour-long conversation, I asked him what the halachic view of the PGP might be—how would it be perceived through the lens of Jewish law? He said he'd never really thought about it. "Are you exposing yourself or others to irrevocable harm that can be avoided? If that's the case, then in the Jewish tradition that's not permissible. But on the other hand, you know the joke from Rabbi Akiva: 'All is foreseen and free will is given.' I don't have an answer. You're never gonna find anything that two Jews agree upon except that some third person should give to charity."[60]

Eventually the IRB agreed to consent to admit two more people but still wanted to proceed with extreme caution. Each of the first ten participants would have to have a master's degree in genetics or equivalent. Presumably those people could be expected to understand the risks of open consent. Or to put it more bluntly, as one bioethicist familiar with the PGP said to me, "If we have highly educated 'altruists' willing to take a hit and potentially go through life without insurance, then we should explore the unintended consequences with *them* because at least they could afford it."[61]

In May 2006 the National Human Genome Research Institute asked for a "single coherent document" that would put forward a scholarly presentation of the PGP and address the attendant ethical, legal, and social issues ("ELSI").[62] George assembled a virtual collection of seventeen bioethics and legal experts (including, I should say, one of my colleagues at Duke, Bob Cook-Deegan, and me). With Dutch doctoral candidate Jeantine Lunshof and Harvard Law student Dan Vorhaus (both have since graduated), George drafted a white paper that tried to make the two cases I articulated above, namely 1) promising genomic privacy is dangerous and unrealistic, and 2) the only way to fully exploit genomic data is by integrating it with other biological and phenotypic data, some of which are intrinsically identifying (faces, for example).

By way of comparison, the white paper discussed—and took a couple of shots at—the consent used in a large-scale global genotyping project of a few years ago, the HapMap, which amounted to a survey of genotypes from four different populations and is widely acknowledged as a very useful resource to researchers who troll through the whole genome in order to find disease genes.[63] The HapMap consent included this passage, which was singled out by George and company:

> Because the database will be public, people who do identity testing, such as for paternity testing or law enforcement, may also use the samples, the database, and the HapMap, to do general research. *However, it will be very hard for anyone to learn anything about you personally from any of this research because none of the samples, the database, or the HapMap will include your name or any other information that could identify you or your family.*[64] [Emphasis in original]

George's original form, on the other hand, said the following:

> You should also be aware of the ways in which knowledge of your genotype and phenotype might be used. For example, anyone with sufficient knowledge could take your

genome and/or posted medical information and use them to (1) infer paternity or other features of your genealogy, (2) claim statistical evidence that could affect your employment or insurance, (3) claim your relatedness to infamous villains, (4) make synthetic DNA and plant it at a crime scene, (5) reveal the possibility of a disease or unknown propensity for a disease.[65]

It read like a list of potential plotlines for the Lifetime Movie Network (which, in 2007, ran a recurring feature titled "'Can You Handle The Truth' Thursday"). The first two scenarios struck me as the most worrisome. Genetic discrimination was an obvious possibility for which we think we have the beginnings of a legal solution. Nonpaternity has been going on since Adam and Eve; we just haven't had a way to confirm it until recently. If you give the same single-gene genetic test to parents (or even one parent) and their/his/her children, and the results don't jibe, there are three possibilities: 1) a new mutation in the child, 2) laboratory error, or 3) nonpaternity. Of the three, assuming one is dealing with a reputable lab, the last is by far the most likely: 10 percent of the time, on average, the nominal father is not the biological father.[66] When I was a genetic-counselor-in-training, we were told never to disclose nonpaternity to anyone but the mother—it was none of our business, it was not why families came to us, and really, who wants to deal with the headaches and heartaches living inside other people's closets. But just as the teenage boy found his biological father in a few short steps, personal genomics has already made it that much simpler for children to find out whether the titular dad sitting at the dinner table shares half—or none—of their genomes. In conducting research for this book, I heard stories of personal genomics companies awkwardly having to negotiate the discovery of nonpaternity.

Public outing of people's genomes is something Boston University's genetic legal scholars Winnie Roche and George Annas have thought about for decades. In the 1980s they agitated for something that goes well beyond the Genetic Information Nondiscrimination Act, which collected dust in Congress for twelve years before becoming law in 2008. They would like to see a statute explicitly guaranteeing genetic privacy.[67] But if there was no sense of urgency to pass GINA, which as I've noted is designed to prevent discrimination only in employment and in health (but not life) insurance, then what is the case for a more far-reaching law?

"Let me just give you a stupid example," said Annas, a short man with glasses and a thick beard who's known for his acid pen. "Let's say you have this technology to do ten genomes a week [cheaply]. That'd become a major part of physicals for political candidates. It would get nasty! Each candidate would want to pull out the genomes of the other candidates and disclose every possible predisposition they have. '*He's* got a pedophilia gene!' '*He's* gonna get Alzheimer's!' 'Oh yeah, well, *he's* probably got Alzheimer's already!' [But] you could certainly draft a criminal statute that says you can't do that."[68] In the wake of the bitter 2008 presidential election, Annas and Boston University neurologist Robert Green wrote a cautionary piece in the *New England Journal of Medicine* citing the possibility of misleading results, false positives, and "genetic McCarthyism" that could result if the electorate started demanding genome scans from presidential candidates.[69]

While George Church included these sorts of caveats in his consent form, it's unlikely he's ever lost a moment's sleep over them. His own behavior has been nothing if not consistent with an unflagging belief in openness. His website offers the best examples of this. To the uninitiated much of it is fairly dry biomedical science stuff: his lab, his funded projects, his publications, his scientific advisory roles.[70] But its many tentacles include pages devoted to him that are not all that different from what a particularly guileless and nerdy college student might post on Facebook or MySpace. His main personal page, for example, features rotating pictures of himself at various ages, describes his upbringing, lists his hobbies (waterskiing [like father like son], aerobatics, turtle breeding, etc.) and charts his development as a scientific thinker from childhood on ("First independent chemistry experiments, Spring 1972. Organic synthesis of cyclohexanone as well as qualitative analysis & IR spectroscopy of various organic compounds"). But it also lets his demographic information all hang out: it includes links to a map of his Brookline neighborhood with an arrow pointing to his house, his signature, his diet and health records, his mother's maiden name, and, until a few years ago, his Social Security number.[71] Even his wife Ting, an otherwise unabashed supporter of her husband, recalled a moment of shock at the latter. "You have your Social Security number open to the *public*?" she said to him. George told her that it didn't matter—if someone wanted it badly enough, they could get it anyway. "Well," she said, "you don't have to make it *that* easy." Down it came.[72]

George's high-profile and sometimes slightly off-the-reservation Web presence, his articles touting the PGP in the media, his rosy predictions about personal genomics on a large scale made to anyone who will listen, and especially his public embrace of the open-source ethos, all served as fodder for those who would characterize him, the way someone did to me with what sounded a lot like suspicion and contempt, as "that guy who wants to put everyone's DNA up on the Internet."

But George has always denied being a dogmatist. The motivation behind his recruitment of a seventeen-strong ethics contingent, he told me, was not to enlist them as a rubber stamp. Rather, he said he did it because he really didn't know exactly what the right thing to do was with respect to subjects' data. (An aside: with all due respect to my social science colleagues, assembling seventeen bioethicists hardly seems like a recipe for consensus.) "We want to stay nimble," George told me not long after we met for the first time. "We are listening and learning."[73]

I asked him how the PGP would handle the discovery of bad news lurking in a subject's genome. What if someone turned out to be at high risk for some catastrophic condition like Huntington's or Lou Gehrig's disease? George said that that person would have the option of *not* knowing.*

---

* Jim Watson, for example, indicated that he did not want to know which version ("allele") of the APOE gene he carries, certain flavors of which could put him at risk for Alzheimer's or heart disease; Craig Venter, on the other hand, is known to carry one high-risk version of APOE.

"Then you're confident you can keep the data private," I said.

"We're not actually confident," he said. "But we are going to try as hard or harder than previous human genetics studies. We're pretty good at computational security and we will try, but we're not promising that. It could become public immediately."[74]

At a genomics meeting I met Margret Hoehe, a raven-haired and out-spoken M.D.-Ph.D. who is a group leader at the Max Planck Institute for Molecular Genetics in Berlin. She recalled getting roundly mocked by colleagues for having the audacity to suggest we begin to sequence humans for medical purposes. This was in 1990. Now in her fifties, she seemed to be no more aligned with the German genetics community—one, admittedly, still wrestling with the colossal eugenic shadow cast by World War II—than she was in her youth. In 2009, the German government passed one of the stricter laws regulating genetic testing in history: all tests must be ordered by a physician, no anonymous paternity testing, no paternity tests during pregnancy except in cases of rape, etc.[75] Whole-genome sequencing of healthy Germans seemed an unlikely prospect. "I'm always twenty years too early," Hoehe said, laughing. She still had big plans for experiments she wanted to carry out, but was not optimistic that she could get sufficient buy-in to land the funding. She described herself as having the soul of an American but having the misfortune of being born German. Without question, she said, the most extraordinary period in her scientific career was her two years in the Church lab in the early 1990s. She stirred her drink and smiled at the memory.

"He's the only one who ever believed in my crazy ideas."[76]

"Why Should We Make Them Go
Out on the Dance Floor?"

3

"**W**hy in God's name would you want to do that?" a geneticist friend asked when I told him I wanted to be a PGP subject and make my genome public. The better question, it seemed to me, was why *wouldn't* I?

At the risk of sounding Pollyannaish, I thought the PGP might actually help make a difference. True believers like Church, Halamka, and some of the other PGP subjects often said that personal genomics represented an opportunity to change health care. Physicians, they said, spend months and hundreds of thousands of dollars trying to find the right statin, the right blood thinner, or the right antidepressant for their patients, to say nothing of the right dose. Trial and error was still the MO. Genomics, they said, had the power to streamline that process: if we knew which genomic variants people carried and the way those influenced drug metabolism, we could prescribe medicines in a more rational way—this is the cornerstone of the burgeoning field of pharmacogenomics.[1] If we knew the contents of folks' genomes, we could avoid giving them the wrong drugs and in doing so save some lives (one study estimated that a hundred thousand people per year suffer fatal adverse drug reactions in hospitals[2]). Further afield, we could potentially use genomic medicine to avoid food allergies, curtail dangerous exercise regimens, and tell people how much sleep is too much . . . or not enough. Prior to the massive Obama health-care overhaul, everyone I spoke to agreed that fundamental changes to health care were essential—in the United States we were spending more than 15 percent of gross domestic product on it and clearly getting ripped off. But should the necessary changes be more about quotidian details like reimbursement and incentives rather than newfangled technologies like supercheap DNA sequencing? Could the PGP help figure out what sorts of useful things one might do with 20,000+ genes appended to people's health records?

I had more selfish hopes, too. I imagined the PGP would be an opportunity to walk in my former research subjects' shoes, albeit with a twist. Scientists who do human subjects research spend so much time writing grants, crafting consent forms, collecting samples, experimenting on and analyzing those samples, and then looking for more, that most of us don't have a clue as to how it feels on the other end of the phlebotomist's needle, the graduate student's personal questions, or the stethoscope. A few years ago I completed a yearlong exercise study, which was fun, got me in shape, and even helped me to drop a few pounds. It came with a few hundred bucks and ten months' membership at a fitness club—a pretty good deal for doing something that I should have been doing anyway. As he biopsied my thigh muscle or pumped me full of glucose, the principal investigator, an amiable cardiologist, patiently answered all of my questions. But at the end of the day, he was not my doctor, he was a *researcher*—his team needed my data infinitely more than it needed what I needed, which was for my cholesterol numbers to improve. I was, finally, just another data point.

The uglier side of this equation can be found in bioethicist Carl El-

liott's brilliant and terrifying 2008 *New Yorker* article, "Guinea Pigging."[3]
In it Elliott described a somewhat freaky underclass of people who eke
out a living as research subjects in clinical drug trials. There were the
subjects who had their recreational time cut short at the investigators'
whims, those in a Philadelphia-based study who were forbidden to leave
their medical research facility on 9/11 ("No one's going home! Everything's fine!"), and most tragically, those in recklessly supervised psychiatric studies who took their own lives.

In part as a hedge against these sorts of outcomes, the initial PGP cohort of ten would be populated by people with biomedical backgrounds
who actually read and understood the consent form and presumably
knew what questions to ask of the principal investigator, who was himself
a subject. The yawning chasm between scientist and subject would not be
so wide. Or so we hoped.

And of course the PGP would also afford me the chance to have a
look at my own genome. After all those years of graduate school and all
those thousands of DNA samples I'd aliquotted into tiny polypropylene
tubes, I might understand something about myself at the molecular level.
Did I carry one or more Hirschsprung's mutations as my nephew must?
Were there some drugs I metabolized better than others? Was I at high
risk for the same heart attacks that felled my paternal grandfather at age
fifty and maternal grandmother at age sixty? Was I in the throes of a
standard-issue midlife crisis or might there be a genetic basis to the anxiety and depression that had made me a regular on a therapist's couch and
at the drive-through pharmacy? Wouldn't it be cool to know?

Accordingly, I sent my slightly alarmed but always practical wife an
improvised email questionnaire. After mocking my formality, Ann responded in earnest (my questions in italics, her answers below them):

> Hi:
>
> *Over the next few days, I'd like for you to think about these*
> *questions and respond in writing. I'm happy to discuss. Thanks—*
> *love you.*

1) What are your overall feelings about the prospect of me getting my genome sequenced mainly for research purposes?

Overall I'm fine with it. I will feel better about it all if I know ahead of time what to expect (in terms of travel and medical interventions) so I can plan.

2) What if my sequence revealed that I'm at high risk for some devastating illness in the future?

Well, that would suck. Can't make policy based on the worst-case scenario. I'm the optimistic type so we'll just hope for the best.

3) What if that same information also suggested that one or both of our daughters are at higher risk for one or more diseases? How would you deal with that?

You mean how would *we* deal with that. I think we'd need to talk with folks about what we could do with this information. I'm thinking here about BRCA; I'm assuming you are, too. Would we have them tested, when would we tell them, etc.*

4) What if my genomic and health information were publicly available on the Web? Would you worry about some person or institution—my employer, an insurance company, a hacker—getting

---

* As I discuss in chapter 10, my mother was diagnosed with early-onset breast cancer at age forty-two—she eventually had two radical mastectomies. She is an Ashkenazi Jew, which means there's a pretty good chance she carries a high-risk mutation in one of the two major familial breast cancer genes, BRCA1 or BRCA2. If so, that means I have a fifty-fifty chance of carrying that mutation while my young daughters each have a 25 percent chance of carrying it. If they carry it, they would likely have an 80 percent lifetime risk of developing breast cancer versus the 15 percent lifetime risk for most American women.

ahold of that information and harming me/us with it? Why or why not?

Again I need to know more to answer this question. What legal protections does an individual have regarding his/her own genetic material? I'm still unclear why, if this information is made public, that the media would need to know your identity. I'm not sure why you cannot remain anonymous.

5) What if getting my genome sequenced meant a lot of press attention? How would you feel about frequent calls, emails and/or visits from reporters?

I'd rather they visit you at Duke. I don't really want the girls involved.

My wife likes her privacy. Unlike her husband, she does not seek the spotlight. Indeed, despite the rise of reality TV and Facebook/Twitter exhibitionism, most Americans still treasure what Justice Louis Brandeis called "the right to be left alone."[4] Thus it was Ann's response to question 4 that seemed to me to get at the core of why the PGP made some people—and the NIH—nervous. The truth of the matter was, outside of a patchwork of state nondiscrimination laws and the Genetic Information Nondiscrimination Act governing health insurance and employment, an individual had limited legal protections over his genome, especially with respect to the physical DNA itself. For most of the last thirty years the judicial system has been fairly consistent: once they leave your body, you do not own your tissue, organs, or bodily fluids or any of the revenue they might generate for a biotech company, a university, or anyone else.[5]

But why? And how did we get here? The issue has still not been completely settled and it's one that continues to reverberate throughout the legal system. In 2003 urologist and prostate cancer researcher Bill Catalona moved from Washington University in St. Louis to Northwestern University after WU tried to assert control over the huge repository of patient samples he had collected. Catalona, a short man with white hair and a slightly gravelly voice, has performed thousands of prostate operations, those of St. Louis Cardinals greats Stan Musial and Joe Torre among them. He knows as much about the clinical genetics of the disease as anyone. Shortly after moving to Northwestern he asked ten thousand patients who were part of his studies to write letters to Washington University requesting that their samples be transferred to Northwestern; six thousand eventually complied.[6] WU responded by filing suit against Catalona.[7]

The case rested on two competing interests: Missouri state property law and federal regulations meant to protect research participants. Catalona had prominent bioethicists and angry patients on his side. Eventually, the Office of Human Research Protections weighed in for Catalona, too. But the courts were having none of it. The Federal District Court ruled that under Missouri law, Washington University owned the samples—the patients had irrevocably "gifted" (such an awful transitive verb) them.[8] If a patient withdrew from the study, it meant only that he needn't donate any *more* samples in the future. Catalona's attorneys argued that the federal regulations covering human subjects, known collectively as the "Common Rule,"* demanded that consent forms never include language that meant subjects had to waive their legal rights. But the Washington University consent forms did indeed include such exculpatory language, characterizing the transfer as a "free and voluntary gift."[9]

The appellate court affirmed the district court's ruling by virtue of what can only be called circular reasoning, saying essentially that federal regulations didn't apply given that the patients, by signing the consent forms, had agreed to freely donate their samples without strings. The court noted that there was no mention of a right to "physically withdraw or request the return of biological samples," nor was there any indication

---

* "Common" because the rules were agreed to by an unprecedented number of government agencies.

that subjects had a right to "direct the transfer of their samples to another entity for research purposes." And because Catalona himself had routinely destroyed samples without consulting subjects, his legal team's "right to opt out" argument was rejected.[10]

Predictably, Washington University crowed about the ruling. The dean of the medical school said it was "of great national importance. . . . This university and others may be unwilling to make those kinds of investments [in biorepositories] if we knew that at any time those collections could be removed."[11] Patients wanting out and asking that their samples be destroyed were told that their samples would instead be "anonymized" and therefore unidentifiable. Once that happened, they were no longer considered "human subjects" and therefore not protected by the Common Rule.

Genetics expert and law professor Lori Andrews of Chicago-Kent College of Law lamented this development and the ruling itself. "Anonymizing a sample . . . is not a satisfactory solution," she told a newspaper in 2007. "Patients should have a right to decide what is done with their tissue. Otherwise it's bait and switch."[12]

Catalona's final appeal—to the U.S. Supreme Court—was denied.

"I think it's terrible," he told me. He noted that anonymizing samples such that they could not be traced back to their donor could have grave clinical consequences. "Somebody may discover a genetic variant that will let us know whether someone in the family is going to get this disease or if his tumor is susceptible to a certain type of drug. And so if a man has prostate cancer and his son has it, it may be very important to compare those tumors to each other."[13]

Catalona has continued to do clinical genetic research in Chicago and still performs three hundred radical prostatectomies a year. But while some patients discussed filing a class-action suit against Washington University, I reckoned that the years of litigation had taken a toll on their champion. "What came out of the case," he said with resignation, "is that the patients could not depend on the court system to follow through on what had been guaranteed to them."[14] More than a year later he was still angry. "The courts have granted Washington University OWNERSHIP of the samples, provided by the patients specifically for prostate cancer research in my program and, as such, WU can now do with them whatever it wishes. My sense is that patients' rights are continuing to erode. Many savvy ones are deciding that the safest strategy is simply not to provide a sample for research."[15]

And what about the rest of us not even involved in research? What claims did anyone have on their own DNA or other body parts? I asked Stanford Law's Hank Greely, who has been contemplating genetics' relationship to the law for decades. A former clerk for Supreme Court Justice Potter Stewart, Greely is a rotund and friendly man with glasses, puffy cheeks, and slits for eyes; when he spoke he reminded me of my in-laws from Ohio (Greely grew up in Columbus), especially the way he pronounced the word *potential* with a long *o*.

"[You have rights to] things that *are* in you," he explained, "not to things that have at one point *been* in you. You spit on a Kleenex and throw it away and, often, you lose. The courts have not yet confronted it, [but] it's come up occasionally in criminal cases where cops follow a suspect around and find a Coke bottle or a beer bottle and get DNA from it for forensic purposes. The reaction has been to call it 'abandoned property.' When you put your trash on the street, the law's clear: You don't own it and anybody can come and get it.

"So you don't have any property-like rights [to your DNA], but you might have a variety of rights to limit or control its use in ways that aren't thought about as property. But all of those would be relatively novel. And they might come from treating it as health information that shouldn't be released or disclosed."[16]

Amy McGuire, a Baylor College of Medicine legal scholar who facilitated the public release of Jim Watson's genome, agreed. "Property law was not designed to handle these kinds of issues."[17]

Greely and I talked about the PGP and he gave it a weak endorsement. "I don't think your position is irrational," he said of my decision to have my genome sequenced and made public. "I'm not an anxious person, but I'd be a little concerned about my children. I have a sixteen-year-old and a nineteen-year-old, and in a sense [if I got sequenced] I'd be revealing half of their genomes. Although the fact we don't know which half helps."[18]

Throughout the writing of this book the subject of anonymity came up again and again. I thought about it often with respect to my daughters. Who did I think I was, taking liberties with *their* genetic information? At an NIH-sponsored Town Hall meeting in December 2006 to discuss how data for large human genomic studies should be handled, one person decried the George Church approach for that very reason: to make one's genome public would be to make family members' genomes public, at least in part.[19] Had I had sons, they would have inherited my Y chromosome almost unchanged (except for a small bit of the X, the Y lacks a dance partner with which to shuffle its genes). As it was, I would be exposing my brothers' Y and my father's Y and my paternal grandfather's Y chromosomes to the world. (That said, the Y is a runt of a chromosome: it has just seventy-eight functional male-specific genes and very few of them are associated with disease.)[20]

But Ann's question was a fair one: Why *couldn't* I remain anonymous (other than to satisfy George's desire for openness and altruism and my need for attention)? A few years ago the *Wall Street Journal* ran a front-page article on a psychotherapy patient who was denied disability coverage for auto-accident-related injuries when her therapist turned over clinical notes to the insurance company. The therapist's notes were mixed in with the patient's general medical records and therefore denied any special protection.[21] I suggested to George that this kind of story—"You can't even trust your shrink to keep your secrets!"—would make "health-information altruism" a tough sell. He seemed to draw the exact opposite conclusion. "It might discourage people who wouldn't or shouldn't be involved anyway," he said, "but it might also convince real altruists that we need to do something fast."[22]

Despite George's "put-it-all-on-the-Web" openness, at one point he expressed misgivings to me that both he and Halamka had outed themselves as subjects before recruiting the rest of the first ten PGPers. I didn't get it—wasn't this the whole point of the PGP exercise? Shouldn't all the subjects be screaming their identities from the rooftops? "I just think it's a team effort and we need to have a little bit of team consideration," he said. "Even if it's something we all plan to do eventually, maybe we all want to do it together, maybe we want to have a press release, maybe we want to make sure that the look and feel of the database are up to snuff before we do it. There are timing issues. There are issues about whether we portray it in a positive light but with enough discussion of the negative aspects. I think that requires some nuance that the really gung-ho people will be too impatient to do. I think this is a work in progress, a successive approximation."[23]

For months I periodically emailed or called George to ask about the status of the PGP recruits: Had they all been picked, had the IRB given them the okay, had they signed the consent form, had they had their blood drawn? The process seemed to be moving at a crawl. I eventually gave up on becoming a subject. George had received hundreds of responses to his call for volunteers. Everyone wanted his or her genome sequenced for free. I imagined that George figured I would write about the project anyway, so why should he even bother to recruit me and risk the possibility that I might write something unflattering? As a science editor I had grown used to dealing with finicky and occasionally prickly scientists fretting over whether writers were capable of getting their work right (the presumption was usually no, they weren't) and whether my portrayals would reflect well on them. I resigned myself to the idea that my role would be to go on masquerading as a journalist. Like Jon Stewart, only not as funny.

One night a message appeared in my inbox. "The Harvard Medical School Institutional Review Board has just approved you as a participant in the study, so if you are still interested . . ."[24] Ha! George really takes this opt-out thing seriously, I thought, though I suspected he was smiling as he typed it.

Ann had just turned out the light and was drifting off to sleep. "I'm in," I announced. "Congratulations, that's great," she said. She was quiet for a moment.

"Or maybe it's not. I guess we'll find out."

When I told friends and colleagues I was officially a PGP subject, after the "Why do this?" question usually came the "Why you?" one. Or, as one said, "What makes *you* so special, Genome Boy?" Nothing. Nothing makes me special: that's the whole idea. (That exchange, however, did prompt me to start a blog called genomeboy.com.)

But in the early days a fair number of scientists and bioethicists perceived personal genomics as special to a fault: in their eyes it was a purely self-indulgent exercise. One bioethicist I know said she thought the sequencing of Watson, Venter, Church, et al. would "help identify the megalomania gene." A May 2007 article in *Nature* was headlined "Celebrity Genomes Alarm Scientists." In it, fly geneticist extraordinaire and onetime Venter nemesis Michael Ashburner called sequencing "famous or very rich people . . . bloody tacky." The Genetics & Public Policy Center's Kathy Hudson said that if this is what came out of the Human Genome Project, it would be "sort of a sad statement." Then–National Human Genome Research Institute director Francis Collins took a veiled (but probably justified) shot at Watson for not consulting his family members before getting sequenced. At the head of the article were smiley pictures of Venter, Watson, and George, the Unholy Trinity of Personal Genomics made to do a perp walk across the pages of science's most hallowed publication.[25]

As a PGP subject I could hardly claim to be objective. But to me this smacked of a double standard. On the one hand, the noble souls at the NIH wanted to protect poor, naïve research subjects and their genomic data; on the other, they seemed to have abiding contempt for the early adopters. If a potential research subject was poor, uneducated, or a minority, then the knee-jerk response was that that person was ripe for exploitation. I get that—the history of genetics is littered with examples of bad behavior toward folks who could not defend themselves, from American eugenics to the Nuremberg laws to Nazi medical experiments to the Tuskegee "studies" of untreated syphilis in African American males that dragged on for forty cruel, inhuman years.[26] But did these outrages necessarily imply the converse to be true? If someone white, educated, and relatively advantaged chose to participate in a cutting-edge experiment, was it fair to presume he or she would reap some enviable benefit or was doing it only for elitist navel-gazing purposes?

I attended the Biology of Genomes meeting at Cold Spring Harbor Laboratory (CSHL) in May 2007. Not so long ago this gathering was called the "Genome Mapping & Sequencing Meeting"; now, in the post-genome era, it was much more about actual biology, that is, how cells work, than about just amassing DNA sequence. The 2007 edition featured lively presentations on Neanderthals, biofuels, and genes that make dogs run faster. There were plenty of famous geneticists around, but the gathering still felt very bottom-up: students and postdocs outnumbered their bosses; they razzed them from the podium and drank with them in the evenings. For me it was nostalgia-inducing. I recalled my days as a real scientist in training, staying up late making posters or PowerPoint slides and then sprinting to catch a train or plane to some exotic locale, making my presentation, and heading to the bar for a long night of revelry interspersed with grad-student talk about genes, postdocs, and life.

When I arrived it was a gorgeous day in early May and everything was already lush in the aftermath of rain and warm temperatures. I saw Chad Nusbaum, sequencing guru from the MIT-Harvard Broad Institute whom I'd met at a few months earlier at the genome technology meeting in Florida, cavorting on the lawn behind the dining hall in shorts, a T-shirt, and bare feet. Chad would often jog the eight miles to work in the Boston snow and ice. Seeing him now I was jealous: in my long-sleeve shirt and sport coat I not only looked every bit the outsider, I was practically panting from the heat as I waded through the lunch line.

The lab—a sprawling campus, actually, that was the unofficial home of the early-twentieth-century eugenics movement[27]—is nestled on a pristine inlet on Long Island Sound. Manhattan is thirty-five miles west but might as well be on the other side of the world. The harbor was dotted with boats and birds; geese squawked overhead. The lucky/well-connected folks got to stay on campus in dorms or the rustic clapboard houses. The rest of us were shuttled to and from the Holiday Inn in a neighboring town.

I was here to try to see Jim Watson, former president of CSHL and, before some unfortunate remarks about race he would make a few months later,[28] president emeritus of the lab. Watson landed here in 1968, the year he published his shockingly impolite memoir *The Double Helix*,[29] and fifteen years after he and Francis Crick discovered the double helix itself, for which they won the Nobel Prize in 1962.[30] I wanted to ask Watson about his decision to get his genome sequenced, how he felt about it, why he'd decided not to learn about his status for APOE (the most important common Alzheimer's susceptibility gene[31]) and what if anything his precedent might mean for the PGP.

Someone at CSHL had made inquiries on my behalf. He reported that "Jim's people" said he was not giving interviews until the end of the month, when there was to be a big event at Baylor University in Houston honoring the completion of his genome. My friend said that maybe I could meet him casually and chat with him but warned me not to push. Watson had long been considered a loose cannon, a trait both he and his handlers recognized (some less recent things he's said: "The best home for a feminist is another person's lab,"[32] organismal biologists are just "stamp collectors," "anyone who would hire an ecologist is out of his mind,"[33] "I've never read the Bible; I'm not sure I've missed much,"[34] etc.). Friends told me that the lab had tried to rein him in over the years but without much success. "You never know what the hell he's going to say," admitted a onetime instructor at a CSHL summer course whose students Watson used to address. She had no idea how prescient she was.

As the meeting went on, he was not around much, save for an outdoor book-signing event for the newest edition of the breezy (no, really) textbook *Recombinant DNA*.[35] I joined the crowd on the veranda just off the bar/café and grudgingly plunked down my credit card (eighty-three dollars . . . for a paperback!) in hopes of gaining a little access. I watched him squinting through his glasses and breathing loudly through his nose, his expression inscrutable. He was engrossed in the act of signing his name, which he did ever so slowly and meticulously in small print, occasionally lifting his head to chat with his coauthors. He barely acknowledged the queue of giddy but polite young scientists. I opened my mouth to say something, only to wimp out and shrink away in authentic fake-journalist fashion. I had been reading about this guy in textbooks for most of my adult life, gazing at his picture, listening to outrageous stories about him. Now, after twenty years, here was my chance and I flinched.

I tried to convince myself that I could craft a composite Watson by talking to the other folks involved in Project Jim. Michael Egholm was vice president of research and development for 454 Life Sciences (bought by Roche in 2007), the company that was first to market with its next-generation DNA sequencer. Project Jim was a million-dollar effort[36] (or maybe $2 million, depending upon whom you asked[37]) initiated by 454 together with Baylor sequencing czar Richard Gibbs, presumably as a way for the company to garner publicity for its new machines.[38] I approached Egholm, an affable if cocky Dane with slicked-back hair, a pink tint to his face, and wire-rimmed glasses. He promised that at the Houston event Watson would be given his genome sequence and that at about the same time there would be a paper published ("hopefully in a good journal") and then, in a symbolic flourish, Watson himself would deposit his sequence into GenBank, the public home of all known freely available DNA sequence information.* The question of what part of his genomic information would be redacted beyond APOE was an open one. "We have a list

---

* A good journal, yes: *Nature* . . but not until April 2008—nearly a year after the announcement of the completion of Watson's genome. A genomics muckety-muck told me the initial data were not of very high quality and the 454/Baylor team was told by reviewers to go back and polish it.

of bad genes," Egholm told me. "We're simply going to walk him through that list and ask him what he wants. The real discussion is, what do his relatives say? The concern is that a year or two from now we're going to discover that if one has this or that allele, one will drop dead at a certain age. Jim doesn't really care about that—he likes to rub that in, you know, that even at his advanced age he doesn't have any maladies. The real concern is about his descendants." Watson had two grown sons, one of whom was schizophrenic.

In practical terms, Egholm said that for next-generation sequencing, data analysis remained the biggest challenge, especially for an entire human genome sequenced six or seven times over. "Twenty billion bases is still twenty billion bases. There's no way around it."[39] (This was in 2007. Six- or seven-fold sequencing coverage [about 20 billion bases] of a human genome would soon be considered laughable. The first sequenced Asian and African genomes, for example, were sequenced at more than 30x coverage, or about 100 billion bases, which became something of an informal standard.)

One difference between Project Jim and the PGP was that the former was not originally conceived of as research at all. Rather, it was an agreement between a company, 454, and a private individual, Watson. When in early 2007 Baylor medico-legal scholar Amy McGuire was brought in to address the ethical aspects of Project Jim, the first thing she did was push to restructure it as a research protocol. This meant drafting a more rigorous consent and getting the Baylor IRB involved. She and I talked at Cold Spring Harbor—if she had any concerns about Watson's persona or that the sequencing of his genome might be perceived as self-serving, she did not betray them.

"I think this is a really unique opportunity," she told me. "He is someone who has extraordinary knowledge and expertise on all of these issues, one of which is: How do we get people to understand the meaning of the data that we're generating? That is a tremendous challenge, especially when we don't understand much of it ourselves. If anybody's going to understand it, it's him. At the same time we could think about the future when these special circumstances are not going to be present. With him we didn't have to worry about the vulnerability aspects."

She was hopeful that Watson and his ilk would allow the bioethics issues surrounding personal genomics to be dealt with beforehand (again, this was before Watson's public meltdown). "I don't like being reactive—it's not fun when the shit hits the fan. It only takes one example to set us back so far—look at the Jesse Gelsinger case."*[40]

I told her she sounded like George.

National Human Genome Research Institute director Francis Collins was at the meeting. I braced myself for the party line about traditional modes of informed consent, and genomics as the fountain of youth and proof of God's egalitarian plan for humanity. (Collins was a born-again evangelical Christian and prone to invoke religious imagery in his pronouncements about the human genome. In 2009, shortly before he was nominated to the directorship of the NIH, he launched the BioLogos Foundation,[41] which "promotes the search for truth in both the natural and spiritual realms, and seeks to harmonize these different perspectives.")[42]

At Cold Spring Harbor, Collins moderated an evening panel on the discovery of genes that appeared to have been selected for in humans during the course of evolution.[43] It was and is a topic raising all of the usual bugaboos about race and ethnicity in genetic and genomic research (and the very topic that would get Watson in deep doo-doo a few months later). Have some human traits been selected for in specific human populations but not others? And the uncomfortable corollary: Are some ethnic groups

---

★ Gelsinger had a rare, genetically caused enzymatic deficiency. He died in 1999 at age eighteen while enrolled in a gene therapy trial at the University of Pennsylvania. The lead researcher was cited by the Food and Drug Administration for flouting several rules. Gelsinger's death is widely acknowledged to have single-handedly derailed the gene therapy field for years.

stronger, faster, smarter than others? It is a rich, controversial, and fascinating question similar to the most provocative ones raised by the PGP: At the biological level, *what are we?* And what do our differences mean?

Despite his casual dress—jeans and blue-striped golf shirt—Collins appeared to be a bit ill at ease. He spoke without identifying specific researchers: "There was a paper published in a very prominent journal suggesting a signature of selection. . . ."

The paper he referred to made claims that a particular gene had been favored over the course of evolution by playing a role in brain size and therefore, it implied, intelligence.[44] Scientifically it was a real stretch, but *Science* ate it up, no doubt because it was such a sexy and controversial story. I listened as Collins went on to exult in the fact that subsequent work showed the connection between this particular gene and intelligence not to be real. Others on the panel chimed in on the paper in question and the delicate business of doing this kind of research. Collins seemed relieved when the subject was changed. At one point during the Q&A he said, "I won't go to IQ because that's probably the most explosive, but let's talk about athletic performance." It seemed to me something of an Orwellian moment: the world's most visible genome scientist felt the need to censor himself in public. He was, in many ways, the anti-Watson.

The next day, between talks I was stalking another übergeneticist when I practically bumped into Collins, who towered over me. I introduced myself and made a lame, passive-aggressive joke about not being able to get through his human firewall of an assistant. I told him I was hanging around George Church, documenting what the PGP was doing. (Was I committing some ethical breach by not disclosing my status as a PGP subject? Probably.) I asked if he would say something on the record about personal genomics. He would not.

I asked him about 454 sequencing Watson, the biotech company Illumina sequencing a Yoruba man, Craig Venter sequencing himself, and George launching the PGP. He was dismissive of all of it. He thought it unfortunate that this was what I was focusing on. He gave me the impression that he thought these high-profile white-guy sequencing efforts were guided by Narcissus. He hoped that this would be a brief, transient chapter in the history of our field. The public, he suggested, would not understand these efforts and would be turned off by them.

When I told him I couldn't speak to what the others were doing but that I believed George's goal was to scale up to thousands or even millions of people and to have the first few people sequenced be those who could sustain any unintended consequences as a means to that end, Collins was unpersuaded. He intimated that it was a self-interested, "look at me" approach to genomics. His bottom line seemed to be that this was an elitist approach, a step in the wrong direction.

The conversation was over.

In the bar/café I saw a man in a kelly green sweater and porkpie hat buying an iced coffee and a huge Danish. Still reeling from my encounter with Collins, I approached feebly and told him I was writing a book on personal genomics and asked him if he would speak to me. "Sure, but I don't have much time," said Watson.

Like Collins, Watson didn't want me to record him, though his demurral was somewhat gentler. He mentioned a magazine article in which he was portrayed as anti-Semitic, which he assured me he was not. In fact, he told me that his father admired the Jews because "they don't believe in God—they use reason." And so he therefore considered himself to be "culturally Jewish." Oy.

We moved to the veranda. Watson leaned against the railing, his face framed by a swaying stand of trees and the harbor in the background. This brought to mind the famous 1953 photograph of Watson and Crick standing with obvious pride in front of their model of DNA. Crick—all sideburns, ears, nose, and especially eyebrows—is standing on the platform upon which the giant helix is fixed, his left hand on his knee, his right gently resting on the sugar-phosphate backbone of the molecule, a small ruler between his fingers. Watson—rumpled shirt, skinny tie, and unruly pompadour—is looking not at the model or Crick or the camera. Rather, his attention is to something somewhere offstage, his smile at once knowing and mischievous. His face appears to be saying, "I can't believe this! And I've only just turned twenty-five!" Standing on the sun-dappled deck with him fifty-four years later, I could have sworn he was wearing the same expression.

When 454 approached him about Project Jim, did he have any hesitation? "Oh no. I never thought twice about it."

When I asked about his reluctance to know his APOE status he said simply that his grandmother died of Alzheimer's at eighty-three and he didn't want to spend the rest of his life worrying that any lapse in memory he might experience would be the onset of dementia. So was he nervous about discovering anything else that might lurk in his genome? "Not at all," he said.

What did he say to Francis Collins about all of this? "I told him how much more reassured I was having had my genome sequenced. And he looked uncomfortable." He laughed, eyes wide.

I asked about the value of sequenced human genomes. "I think we will have a more compassionate, better society because of them," he said. "They will be useful to explain why people are the way they are, why some people don't fit in. My son has schizophrenia, for example. He never wanted to go to work or to school because he couldn't fit in. Maybe genomics can explain that. We are all different. We all have different abilities. Some people can't sing or dance. Why should we make them go out on the dance floor? I think it's better to understand people's limitations. If we understand the biological basis of those limitations, then I believe we will treat people better."

Did he get permission from his children? "No," he said. "They might have said no."[45] He laughed again. I recalled George's lament that because of Watson's bull-in-a-china-shop approach, perhaps Watson was not a good choice to start with for personal genomics.[46]

I asked about the concerns some people had about informed consent and individual whole-genome sequences. He was unmoved. "If we get ourselves wrapped up in this, then no information will *ever* be released." And what about charges of elitism? "It's crap! Not everyone can afford an automobile. Does that mean the rest of us shouldn't be allowed to drive? People who want these types of restrictions on genetics *don't like genetics*. They see genetic determinism everywhere and they think it's an evil worse than fascism. No competent geneticist can believe in genetic determinism."[47]

I regretted not asking him why, then, if he were not a genetic determinist, he would be so discomfited by knowing his status for APOE, which, after all, is only a susceptibility gene. While it alters one's risk for Alzheimer's, in its worst guise it is still well short of a guaranteed death sentence. (Persons carrying two copies of the APOE4 allele have a fifteen-fold increased risk for developing Alzheimer's.)

Every time Watson flitted away and I thought the discussion was over, he drifted back with an addendum. The conversation caromed from Michael Crichton ("He told me he was on his fifth wife—I thought that explained a lot") to religion ("I don't hate God the way Francis Crick did") to psychiatrists ("They don't know very much").

Finally he signaled he was really leaving, but not before imparting some literary advice. "I wouldn't spend more than twenty-five pages of your book on George Church," he said. "He's not that interesting . . . though he's very tall. He can be one character . . . but you need more. You need a villain if you're going to compete with the Michael Crichtons of the world. Who's your villain?"

I hesitated. "Gosh, I don't know," I said.

"You *have* to have a villain," he insisted.

"Well . . . right now I suppose it's Francis," I said without much conviction.

"Aha! That's good. So now you've got at least two thousand readers."[48] I looked around nervously—there were chortles of affirmation from some of the other folks who'd gathered on the veranda and were eavesdropping if only to hear what outrageous thing Watson might say next. One scientist assured me that if Francis Collins were indeed my bad guy then she would be among the two thousand. I started to talk with her; when I looked up Watson was gone.

Later that afternoon, the organizers of the meeting grudgingly gave him a forum to talk about his genome. The feeling seemed to be that this was a nonevent, though I wasn't quite sure why. Maybe because human genomes were still not taken seriously—they weren't "real" science; rather, they were thought of as tiny blips of data that added up to nothing. Maybe because Watson was likely to say something outrageous (a pretty safe bet). Maybe because it had simply become too much work for the veterans of these meetings to manage and minister to him. A longtime CSHL denizen had seen this tug-of-war before and offered a note of sympathy for Watson. "It's hard being ex-emperor when people have been saying yes to you for sixty years."

Watson made it clear that his genome—and all human genomes—should be cause for celebration. Having them would make us better, healthier people. "If there had been a genetic diagnosis for my son Rufus," Watson told the audience, "we would have raised him differently and not expected him to go to Exeter. . . . I'm not hesitant to say we're playing God. Someone should."[49]

Ting Wu gave a congenial laugh at the notion of her husband as an elitist. "I think when people get to know him that thought will vanish. He's very modest. People won't be able to miss it."[50]

When I raised the issue with George, he patiently explained—again—that we ten were not really important. We would be among the first, sure, but he reminded me that his premium was on scalability, that is, getting to a million. Thus he opted to sequence only our protein-coding 1 percent, our exomes—at least for now. "If you give me a hundred million dollars," he told me, "I would rather spend it on ten thousand or one hundred thousand exomes than on one hundred complete genomes."[51] In other words, he would rather have 1 percent of the genome from one hundred thousand people than 100 percent of the genome from a hundred people. To paraphrase *The Incredibles*, once everyone's genome was super, then no one's would be.

But to scale up meant moving beyond the limited pool of master's-level geneticists willing to share their genomic information (there are only a few hundred board-certified medical geneticists in the United States and only about 2,500 master's-level genetic counselors).[52] George therefore wanted to replace the credential requirement with an exam that all PGP subjects from any walk of life would have to pass before being allowed to enroll. "A test could help prove they know what they're doing and might even have a psychosocial component to make sure they're not pathologically exhibitionistic or likely to react in a depressed way to news about, say, Huntington's disease." What else would it cover? "Interactions with insurance companies. Things like 'Do we think having reporters in our living room is a good thing or a bad thing?' And 'How much do we think our cells are worth?'"[53]

George saw money less as something that would taint and commoditize personal genomics than as a means of altering the subject-researcher power dynamic. Indeed, he suggested he might be willing to blur the scientist-subject distinction altogether. While some bioethicists have argued that returning genetic research results to subjects is a good idea, it's not clear how many feel that way. (Some have issued "consensus" statements,[54] but as far as I can tell, while putting out consensus statements is a favorite pastime for bioethicists, it tends to correlate poorly with actual consensus.) For their part, all the data say that participants would very much like their results back.[55]

Certainly there are good reasons *not* to give experimental results back: research is just research, after all—it has not been clinically validated. Do the researchers even know what the data mean, especially if they're not clinicians? And what if the information they give is wrong—can they be counted on to clean up the fallout from results that mislead or, more likely, just don't pan out?

George didn't buy it: he had long thought disclosure of genetic and genomic results to subjects was the natural thing to do. He said he would like to make PGP subjects true partners in the outcome of his project. Yes, they would bear the risk of eating from the tree of genomic knowledge, however bitter and uncertain that might taste, but they would also reap the rewards. "We're hoping everyone in the study will not only know what's going on, they will actually be working with us to analyze their own data. You could be so informed you might even qualify as a coauthor."[56]

He said that returning results to subjects was imperative because it would give them the information they needed to make the single most important decision about the PGP that other human genetics protocols did not: when to quit. "I think the opt-out clause of most consent forms is a mirage," he told me. "It's a fake. If you don't get your information back, then how would you know you have a Huntington's mutation and don't want that to be in the public domain or even a private database? Ideally, opting out means that that type of information would be erased from everything. But if you don't know it to begin with, then you can't ask the investigators to erase it."[57]

"So what will you do with your exome?" I asked him in early 2007.

"Um . . . what do you mean? I'm going to *study* it."

"Okay. But will *I* be able to study it?" I felt like a six-year-old boy at summer camp: I'll show you mine if you show me yours.

He demurred and suddenly sounded more like the modest, private man Ting described to me. "I'm not going to be superfast in putting mine in the public domain. I intend to, but I think I would like to be a guinea pig for the phasing process where I'll look at it myself as much as I can with software, and then get some experts involved who are inside the PGP and look at it with them with moderate security. And then together if the PGP subjects and the PGP researchers feel that there's anything that needs to be redacted, then we will redact it either from the genotype or the phenotype or both. We will leave behind a scar that says 'this was redacted.' We won't say, 'Oh no, George doesn't have schizophrenia.' We will say, 'We're not *saying* whether George has schizophrenia or not.' And those scars will be revealing in a certain sense, but they may not be actionable. We'll work with ELSI scholars and genetic counselors and genetic experts to try to proactively redact, and not just for me but for everybody else in the PGP who wants to do this."[58]

George told me his daughter wanted her personal genome—or some approximation of it, anyway—for her sixteenth birthday; he and Ting had tentatively agreed. I met her at the first PGP barbecue at the Church-Wu home. She was a tall, striking teenager with her mother's jet-black hair and her father's liquid, penetrating eyes. And while she didn't like school ("not my thing"), she seemed to share her parents' propensity for creativity and overachievement. She is an artist: her abstract photos, paintings, and multimedia creations adorned the walls of the house. She told me the plate I was eating off of was one of hers. She designs clothes. And she's a fashion model ("I'm trying to sign with a big agency").

She shared three short-term goals. "On my birthday I'm gonna go to school and tell them I'm not coming anymore, I'm gonna get a tattoo, and I'm gonna get my genome done." Why did she want her genome? "I'd really like to know what's coming. If I'm gonna have a short life or if there will be uncomfortable things in my future, I really wanna live now." This was the flip side of Jim Watson: he needed *not* to know at the end of his life, she needed *to* know at the beginning of hers. They seemed to share a fatalistic, deterministic view of their genomes.

I asked if she thought she was unique or whether her friends were thinking about their personal genomes, too. She assured me they were, but qualified her answer like a true scientist. "This is Brookline," she said. "Harvard is right down the street. So this isn't a random sample of teenagers. A lot of my friends know about it because of my dad and also because their own parents are aware of it."

"How much of your and your friends' interest in personal genomics is related to MySpace and Facebook—people just putting everything on the Internet?"

"I really don't think any of it has to do with that," she said. "I just think we're a more informed generation because of the Internet in general. Anyway, MySpace is out now—it's all old men and creepy people."[59]

"Playing in bad rock bands," I offered.

"Pretty much."

George said that five thousand dollars for a set of genotypes (the projected price for a scan of a half million genetic markers in mid-2007) was not much of a lifetime investment to make on behalf of a teenager. And pretty soon, everyone would be doing it.

Even before the recent explosion of personal genomics, there were signs that the ossified, neglected, backwater specialty of medical genetics was about to change. Jason Bobe started the blog The Personal Genome: Genomics as a Medical Tool and Lifestyle Choice in 2003.[60] In early 2007, just before taking a job helping George to manage the PGP, he told me he was planning to launch a new website, GenomeHacker.com, designed to give young people crude ways of interrogating their genomes without the benefit of fancy lab equipment. He expected they would be able to infer things about their DNA simply by studying their own phenotypes; for example, if they couldn't drink coffee late in the day without being up all night, it was likely they were slow metabolizers of caffeine and therefore harbored a genetic variant in one of the major genes that encodes a drug-metabolizing enzyme. "Ten years from now there's going to be a whole bunch of fourteen-year-olds developing tools for parsing their genomes," said Bobe. "It'll be a new after-school hobby for kids, I imagine."[61]

Bobe, a curious and precocious guy who turned thirty in 2009 and has become a friend, was enthusiastic about technology and life as depicted in the pages of *Wired*. He liked to send emails at 4 A.M.; his Gmail status message often revealed how many unanswered emails were currently in his inbox (two hundred to four hundred seemed to be the norm). He was stocky and still spoke with a midwestern twang; he credited his blog with rescuing him from the Indiana cornfields.

I never doubted the viability of his idea: genome hacking as an after-school endeavor. But it had already become clear that it was not going to take ten years. Or five. Or even one. Neither kids nor their parents would have to infer their genotypes from their phenotypes. They could go right to the source.

4

"But When Is It Mature?"

When I met Matt Crenson on an overcast Wednesday morning, his deep, raspy voice belied an affable manner, which was appropriate: his employer, the Mountain View, California–based company 23andMe, had for the last several months taken great pains to portray itself as friendly and nonthreatening—the anti-*Gattaca*. George Church, who was on the company's scientific advisory board (he was on some eighteen other such boards, too; God knows how he found the time), often described 23andMe as "playful." The lobby of the company's home in a nondescript Silicon Valley office building looked something like a hipster toy store that was building up its inventory before officially opening its doors for business.

The atmosphere—a basal bustle of twenty-somethings, occasional peals of laughter echoing among the cubes and glass offices—harked back to the dot-com era: there were shelves filled with Day-Glo-colored squishy rubber balls while the otherwise Spartan lobby was scattered about with cute stickers, buttons, and plastic packages of Mike & Ike's red and green candies festooned with the company logo and its catchphrase, "Genetics Just Got Personal." (I stuffed several in my bag for the plane ride home.) 23andMe thought of itself not as a health-care company or as a biotech, but rather as an Internet start-up. "Web 2.0" was the descriptive phrase one heard over and over from its employees.

23andMe officially launched in November 2007, just a couple of days after the first commercial entrant into the personal genomics market, deCODEme, which was an outgrowth of Icelandic genetics-meets-pharma firm deCODE Genetics. Both companies began by offering customers access to what I will call "the variome." If your genome is all 6 billion DNA base pairs (the function of most of which we don't understand), and your exome is the 20,000+ genes (about 60 million base pairs) that code for protein, then your variome is a smaller subset still: it is an assortment of markers more or less evenly spaced across the genome that tend to vary from person to person; some markers fall within genes, but most do not. By early 2010 researchers had identified nearly 13 million of these markers; from 2007 to 2009 the companies typically typed 500,000 to 1 million of them (about 1–2 million base pairs total). These marker sets (called single nucleotide polymorphisms, or SNPs—"snips") were thought to capture much of the variation in human DNA, although they represented no more than 0.05 percent of the entire genome.* But they were relatively cheap to type given that such endeavors would have cost hundreds of thousands of dollars in the 1990s; both 23andMe and deCODEme began by charging customers a thousand dollars. Custom-

---

* Another source of difference can be found in much bigger stretches of DNA that vary in how many times they are present in an individual. You, for example, may carry five copies of a million-base-pair stretch of DNA on chromosome 17 while your next-door neighbor may have seven copies of the same region. These recently discovered bits are known as copy-number variants (CNVs). In 2008, the personal genomics company Navigenics typed customers for approximately 1 million CNVs.

ers entered their credit card number online, waited for their kits, spit in a tube, put it in the mail, and a few weeks later could log onto a secure website and look at their variomes.

Crenson used to be a science reporter for the Associated Press. Given the downward trajectory of the newspaper business, he seemed genuinely happy to land on his feet at 23andMe as "Content Manager," the guy responsible for editing everything the customer read. This particular morning, however, the customer account he tried to show me, belonging to the whimsically named "Greg Mendel,"* wasn't working. After some tinkering and technical help from a young guy dressed in black, we finally got to meet Greg Mendel's genome. Matt walked me through it and pointed out various risk alleles for prostate cancer. "This one raises his risk to 1.29, but this one has only been studied in African Americans. There's a lot of uncertainty now about how much ethnicity affects the results."†

Crenson explained the company's criteria for including a trait among the company's officially sanctioned list, what it called the "Gene Journal." To make the Gene Journal, a trait had to have been studied in a thousand people or more, been independently replicated in another study, and been reported in a reputable, peer-reviewed journal. "Not the *Albanian Journal of Medical Genetics and Dentistry*," Crenson assured me.[1] In December 2007 the company featured eighteen traits in the Gene Journal, from

---

* "Greg Mendel" was later "outed" as 23andMe cofounder Linda Avey's husband.
† This was true—even the companies themselves disagreed as to the extent that race and ethnicity had an impact on individual risks. deCODEme tended to discount ethnicity while Navigenics viewed it as very important. Why does this matter? Genetics is all about context; genes do not operate in a vacuum. You and I, for example, may both have the same gene variant that affects the way we metabolize vitamin D. But if I live in Greenland and you live in sunny West Africa, that variant may have much different effects on our resistance to melanoma, our skin pigmentation, etc. Furthermore, if your ancestors have lived somewhere for thousands of years, you have likely inherited a whole mess of gene variants that are of particular relevance to survival in the local climate.

earwax consistency to restless legs syndrome to Crohn's disease. Within a few months there would be a total of seventy-eight traits. Soon it was no longer called the Gene Journal, but simply "Health and Traits."[2] By 2010 the company was reporting risk estimates on forty-seven "clinical" traits that it considered to be fully vetted (including twenty-one recessive diseases for which one might be a carrier), and eighty-seven additional "research" traits for which either there was insufficient data (in 23and-Me's eyes) or the traits did not affect disease risk (for example, "food preference," "hair color").[3]

From the beginning 23andMe offered customers information about their ancestry using mitochondrial DNA (passed on only by mothers) and Y-chromosome markers (passed on only by fathers); eventually it began including markers from the other twenty-three chromosomes (1 through 22 plus X). Crenson took me back in time through "Mendel's" geneal-ogy and the groups of markers that had been transmitted down through his family. A few months after my first visit, 23andMe would incorpo-rate social networking based on genomic characteristics into its menu: a Facebook for the genome-savvy set. At the Global Economic Forum in Davos, Switzerland, company executives passed out spit kits to the glit-terati, an apparent strategy to help brand personal genomics as hip.[4] Bono was tested. Jimmy Buffett and Warren Buffett were tested: no relation, they learned![5] Peter Gabriel was tested.[6] In 2009, 23andMe launched an online community of pregnant "mommy bloggers": "Explore the genetic legacy your child will inherit from you and your partner."[7]

23andMe was cofounded by Anne Wojcicki, the wife of Google zil-lionaire Sergey Brin, and Linda Avey, formerly of biotech behemoth Af-fymetrix and its now-defunct human genomics spinoff, Perlegen. On the day I visited, true to 23andMe's start-up mojo, both women were running around with great urgency, putting out fires and hurrying to and from meetings and conference calls. They apologized and asked if I could come back later. That afternoon, Avey, a striking blonde in her late forties who grew up in South Dakota, finally sat down to talk about the origins of the company.

When Avey was at Perlegen, her mission was to convince pharmaceutical companies to use the Affymetrix GeneChip technology—the small glass wafers used to type some of the millions of SNPs in human and other genomes—to begin to go after genes that cause specific diseases. But in those days the genome was terra that was even more incognita than it is now—relatively few SNPs had been characterized. Thus Perlegen couldn't begin gene discovery without first embarking on an expedition to isolate more markers that would allow it to sharpen the cartography of the genome. That meant a $100 million commitment to sequence bits and pieces of fifty human genomes, however crudely.[8] This was an arduous slog, but it paid off: the company managed to identify 1.7 million SNPs, an unprecedented treasure trove in 2003.[9] (For comparison, when I began graduate school in the early 1990s we knew about no more than a few hundred polymorphic DNA markers across the entire human genome.)

Perlegen was now in a position to design a DNA chip with several hundred thousand markers and begin to do genome-wide association studies (GWAS). These are essentially very dense case-control studies designed to find DNA markers important in disease. By typing the same set of markers in large numbers of cases and controls, it becomes a brute-force statistical matter of finding markers that pop up more frequently in sick people than in healthy ones. Those markers are very likely to be in or near genes that play a role in disease.[10]

GWAS studies have since become ho-hum.* But only a few years ago, they were the new new thing. So new, in fact, that NIH and industry were reluctant to fund them. By 2005, Perlegen had designed a chip with 250,000 markers, identified cohorts with diseases such as Parkinson's, but could get neither corporate backing nor public funding to move ahead with GWAS.[11]

---

* In many ways, GWAS have been a disappointment: we've found a lot of disease genes, but they are weak—they don't explain very much of the various diseases and that makes it hard to use that information in the clinic. If Crohn's disease is caused by twenty genes, how do we design a drug that targets twenty proteins?

"You might identify a group of two thousand samples," Avey said, "but the NIH funding mechanisms were stuck in their old paradigm— they had cutoffs. People would look at their sample sets and say, 'Okay, [given my funding] I can only afford to do a certain number.' And then it would sort of defeat the purpose because you wouldn't have the statistical power to really find what you were looking for. The whole vicious cycle was infuriating after a while. A lot of people at Perlegen and Affy were frustrated. This technology was just sitting there and being underutilized."[12]

It was this frustration, Avey told me, that planted the seed for a freestanding personal genomics company. "What if," she thought, "we just shifted this paradigm and opened up the ability for people to pay their own way and gave them access to their own genetic information?"[13]

Her first idea was spas. Spas, she reasoned, had become more clinically oriented while doctors' offices had become increasingly more spalike. "You had this weird convergence happening. People at spas have got disposable income and they are very interested in their health. Some of the high-end spas had changed their image to become 'wellness centers.'"[14]

Avey took the concept to some of the Affy VPs with whom she went on six-mile runs in the morning. "Talk to Steve," they told her. According to Avey, Affy CEO Steve Fodor was enthusiastic and told Avey she had to do it . . . but not at Affy. It was, he told her, beyond the firm's core mission as a research tool provider.

At a meeting where Perlegen was presenting data on the company's efforts in Parkinson's disease, Avey met Google mogul Sergey Brin, whose mother suffered from Parkinson's.[15] Brin began asking Avey detailed questions about Perlegen and its analytic approach. Hopeful that she could persuade Google to back her new venture, Avey began trying to set up meetings with Brin, but found it difficult.

"He always wanted to have his girlfriend there with him. And I thought, you know, who is this girlfriend?"[16]

Anne Wojcicki had been investing in health care and was herself frustrated at the sector's lack of progress and poor return on investment. She and

Avey came together at the annual Technology, Entertainment and Design meeting in Monterey and agreed to move ahead with what they called— aptly if unimaginatively—"Newco." Google would kick in $3.9 million and Newco would eventually become 23andMe. Wojcicki told *Fast Company* that she was in her kitchen reading Wikipedia, saw a picture of the twenty- three pairs of chromosomes, and started singing the words, "Twenty-three and me." Newco had a new name.[17]

Where 23andMe went for playful, its Silicon Valley neighbor, Navigen- ics, opted for serious. Serious as a heart attack, one might say. Whereas the 2009 iteration of the 23andMe Web presence was all light and shiny and iPod nano–ish, the Navigenics website felt more like a doctor's of- fice, full of pastel and sepia tones, handsome and mature couples, all sun-swept and looking extremely healthy—one half expected a medi- cal history questionnaire and six-month-old magazines to materialize on one's screen. The original video I saw describing the company's Health Compass service was presented by cofounder and chief scientific officer Dietrich Stephan, a short, amiable, and balding man in his early forties. In the video he appeared to me to be somewhat uncomfortable and con- stricted; in person he reminded me of a gentler, more garrulous, and perhaps more rumpled version of the great B-movie actor John Saxon. When I first met him at company headquarters he wore baggy pants and layers—shirt, red fleece, and brown cotton jacket. He spoke softly and with a faint Pittsburgh accent; he laughed easily, even when talking about the minutiae of FDA regulations.

He began plodding along from square one, giving me a rudimentary history of the Human Genome Project and Craig Venter. I tried not to make my impatience too obvious but rather to politely move him beyond the scripted VC speech. When I asked him about the impetus for starting Navigenics, he became animated, just as Linda Avey did when talking about the birth of 23andMe. Stephan was, he told me, toiling away at a large genomics research shop in Arizona, the Translational Genomics Research Institute (TGen), finding disease susceptibility genes and watching his discoveries go . . . nowhere. "I started getting frustrated that people weren't applying this to understanding individual risk profiles. I tried desperately to get things placed into molecular diagnostic facilities, but they didn't understand how to interpret that information and give it back to physicians or genetic counselors. They would say, 'This is not a binary diagnostic.' That's what all of these folks are used to: you have two copies of a cystic fibrosis mutation and you're going to have cystic fibrosis. Or you have one copy and you're a carrier. Or you have zero copies [and you're neither]."[18]

The diseases Navigenics focused on initially—type 2 diabetes, Crohn's disease, heart disease, multiple sclerosis, obesity, rheumatoid arthritis, and a dozen others—rarely played by Mendel's simple rules of genetics, that is, one gene = one trait. One can carry half a dozen versions (risk alleles) from six different genes that predispose to a common disease and still be at average or even below average risk for developing that disease. Stephan immediately recognized the challenges Navigenics would face. "How do you put that information in context and how do you communicate risk to people?"[19]

When describing the field, he used the word *nuanced* several times. He spoke about educating physicians and genetic counselors and developing a gold standard. "We want to set the bar high. We're going to come at this from a hard-core medicine and science perspective, put all of the spokes of this wheel in place and then roll it out."[20]

One of the early criticisms of companies like Navigenics was that they were giving information about diseases for which nothing could be done. I asked Stephan for an example of a condition that was both multigenic and "actionable." Without hesitation he said, "Alzheimer's."

"It's all about finding people four or five decades before they show the first symptoms, lowering their cholesterol, making sure they live an active physical and mental lifestyle, and taking advantage of some of the early screening programs. And if you do that, you've decreased the incidence by fifty percent and you've prevented ten million people from getting Alzheimer's disease."[21]

From many scientists this type of talk might have sounded like blather—hubris mixed with naïveté. Even though what we know about the biology of Alzheimer's could fill volumes, preventing millions from succumbing to the disease seemed quixotic at best, a delusional and destructive fairy tale at worst. Did Stephan have a messiah complex? He went on about how this model could be applied to other common chronic diseases, such as cancer. The money we would save would help solve the U.S. health-care crisis. He believed it. And so did the investment community. In eighteen months Navigenics had enlisted VC big boys Kleiner Perkins and Sequoia Capital to the tune of $25 million,[22] a tidy sum in a tough market, especially for a company that was not making widgets, but trafficking purely in information.

Stephan brought me into a room full of geneticists, their desks lining the perimeter of an office that looked out onto a man-made lagoon. They were young and enthusiastic, if a bit stressed—good old-fashioned dot-com-style fatigue. "We should have bars on the windows," Dietrich said.[23]

He introduced me to Eran Halperin, a friendly Israeli and at the time the company's head of genetic research; unshaven, he looked tired. A computational guy by training, he had been working hard on finding statistical ways to infer people's genotypes at places where the million markers they used failed to provide information—the genes of interest had slipped through the cracks and were in places in the genome that were not represented by any of these markers. But because chunks of chromosomes tend to travel together as they're passed from parent to child, it was possible to play a kind of probabilistic guessing game known as imputation that could often fill in the gaps between the mile markers represented by the SNPs.[24] Halperin's unofficial title was the Imputator.

Across the room sat Elissa Levin, an attractive, kinky-haired, and gregarious genetic counselor. She directed the Navigenics Genetic Counseling Program and I could recognize her genetic counselor persona—both warm and direct. Part of "getting it right," Stephan had told me, would be providing customers with phone counseling after they received their results. But why? Weren't most of these folks going to be highly motivated early adopters anyway?

"This is a lot of information," said Levin. "We're doing everything we can to try to present it in an eloquent visual and written way. But we understand that it's also critical for someone to be able to pick up the phone. It's not just calling your doctor, who might not be familiar with it. It's having somebody on the other end of the line who can actually review your results with you and understand the context."[25]

In the coming months, deCODEme would make referrals to genetic counselors part of its services.[26] 23andMe included a pointer to the American College of Medical Genetics on its website.[27] DNA Direct, which served as a clearinghouse for traditional single-gene disease testing, employed several genetic counselors,[28] as did genome scanning latecomer Pathway Genomics.[29] But in the early days, Navigenics remained the lone consumer genome-scanning company with genetic counselors on staff that included counseling in its fee.

The company hosted a two-week launch party in New York, at a loft in SoHo. Outside flew a Navigenics flag. In the windows were blowups of the same people I'd seen on the website: the healthy-and-happy-but-serious people of different races and genders and sexual preferences. No fat people, though. Inside there were high ceilings and hardwood floors and generous bags of swag. There was an open bar serving, among other things, a delectable concoction that looked exactly like Hawaiian Punch called a Navatini. Venture capitalists know how to throw a party.

Beautiful people abounded. Al Gore, his climate-change star still at its apex, held forth at the initial reception. "This is a great firm," he said. "On all these new genetic breakthroughs, there is always some resistance culturally, and then, when there's an evaluation of the inherent value, if the ethics are right, if the surrounding culture is right, then it just breaks through. . . . I think this company has the culture right . . . and I think it's going to be a fantastic success."[30]

The buzz was palpable. The *New York Times* style section had a splashy feature.[31] At the loft each day, the Navatini-induced revelry was mixed with gravitas as panels of mostly hot-shit scientists convened to discuss Important Topics relating to genomics and health care. An old friend from graduate school, Tara Matise, a statistical geneticist at Rutgers, had emailed me weeks earlier asking about 23andMe on behalf of her father, who was interested in his risks for Alzheimer's. I mentioned Navigenics and, after catching wind of the party, she agreed to meet me there. "It's possible it's a little on the early side," she said. "I think people like me are curious but they're concerned [that it's not ready]."[32]

We witnessed those concerns firsthand. Any notions that this would be a Tax-Day Love Fest were dispelled by the session innocuously titled "Genomic Testing and Your Practice." It featured Elissa, Dietrich, and two other genetic counselors on a makeshift white stage with an embedded video screen in the back room of the SoHo space. The audience was primarily genetic counselors, all women, and from the beginning their collective stance was skeptical . . . to say the least.

As I've said, in the late 1980s I was a board-eligible genetic counselor. I was never very good at delivering bad news and the moment I fell in love—or at least lust—with lab-based genetic research, I gave up most of my thoughts of working in a genetics clinic and promptly bluffed my way into a Ph.D. program. But I have retained a soft spot for genetic counseling as a profession and genetic counselors as people. Like nurses, I think they have chosen a fascinating and invaluable but quite impossible and largely thankless profession. They have to deal with stressed-out and/or grieving patients on the one hand and often overbearing and capricious doctors on the other. Because with few exceptions they have only master's degrees, they are regularly dismissed and patronized by physicians, especially those physicians who aren't geneticists and therefore don't—or can't—appreciate what counselors do. I've had more than one M.D. geneticist tell me that counselors are no more than "physician extenders." Would they call them that, I wondered, if their ranks were 94 percent men instead of 94 percent women?[33] Genetic counselors were underpaid, despite a common refrain that there weren't enough of them to go around. Like nurses, they lacked autonomy. For decades they were told to be "nondirective" lest they unwittingly let slip their views about matters such as abortion with their patients. Have doctors ever been subjected to such rules?

But on this night in SoHo, the sisters were doing it for themselves. During the Q&A the indignation that had been building found its voice. "First of all this is a class issue," said one counselor from an Alzheimer's clinic, presumably referring to the company's $2,500 price tag and how it would put Navigenics' service beyond the reach of many. "Secondly, this runs counter to everything we are taught as genetics people. Family history is the gold standard. . . . And finally, I don't see how [Alzheimer's] is preventable . . . How does this particular disease fit into your criteria?"

The cognitive dissonance between the newfangled personal genomics model and old-school genetics continued to reverberate among the counselors as I drifted through the crowd and eavesdropped:

- "What happened to the standards we were taught in school? Sensitivity, specificity, positive predictive value, having an intervention?"
- "They're clinically applying this information that's not been replicated, the studies are underpowered . . . I just don't get it."
- "It's cutting edge and exciting . . . but I feel like it's premature."

"What's everybody reading in the [*New York Times*] Science Times every Tuesday?" said Deb McDermott, a counselor based at Weill Cornell Medical College in New York. "They're worried about cancer, Alzheimer's, heart attack. We still don't know why eighty-five percent of women with BRCA mutations get breast cancer and fifteen percent don't. Did they eat an apple every day? Did they smoke or not smoke? Did they take oral contraceptives? No one can tell me because no one's done the studies. So it's bullshit to think that testing a sixty-five-year-old Wall Street banker, who can afford this test, who can start going to his personal trainer and going to a nutritionist, will prevent him from getting, say, prostate cancer. It's bullshit!"

"Would you feel differently if it were fifty bucks?" I asked.

"No. [This company is] not in Indianapolis or Columbus. They're in New York for a reason. It's all about money. We're a couple of subway stops up from Wall Street."

"And if I came to you with these results," I said, "and they showed I had an elevated risk of heart attack, what would you do?"

"I would throw the test results out and say, 'Let's look at *you* and how *you've* lived your life. Let's look at your family history. And I'm sorry you didn't spend that twenty-five hundred bucks on a trip to Jamaica.' These are population data they're trying to apply to single individuals—it just doesn't work."[34]

A few months earlier a scientist friend who didn't want her name used said almost the same thing verbatim. "Dietrich is a nice guy. The 23andMe people are nice ladies. But I don't know what this is going to do for anybody. It's a waste of time and money."

The next day I met with Dietrich in the cavernous basement of the loft, which resembled a theatrical costume studio crossed with a junk shop. He and I sat in a set of mod 1960s chairs beneath the winding staircase. Dietrich had a cold and sniffled throughout our conversation; he had been in New York for two weeks and looked as though he hadn't slept since arriving. Was he put off by the counselors' hostility? I asked. Had he expected it?

"I didn't expect it. But I was more concerned that some genetic counselors didn't seem to understand that family history is really just a surrogate for exact testing of the genetics of a person. Once you have the ability to extract all of the genetic information using a sequence, the need for a family history will go away, I would think."

He thought that perhaps the counselors felt threatened but said that they needn't have: traditional counseling for single-gene disorders would always be around. But we were now in a position to get access to our complex, common genetic risk factors, not just rare genetic disorders controlled by a single strong ("highly penetrant") gene. "I think the counselors will have to learn that those are two different things."

He was equally sanguine about criticism of Navigenics' pricing. "I'm not worried about that at all. There's not much we can do about it. It is what it is. This is a market-driven economy. If you don't like that, go move to a communist country."[35]

These words appear much harsher on the page than they sounded in person. Dietrich may not have shared much in common with George Church, but he was at least as comfortable in his own skin. He never raised his voice and was a consummate diplomat, someone who could tell you to go to hell (or to a communist country) and somehow you would find yourself looking forward to the journey. Despite his illness and exhaustion, despite the undisguised unfriendliness of the medical establishment toward Navigenics and its competitors, he seemed genuinely happy with the events of the previous ten days. Yes, it was a pure and shameless marketing ploy: go to the heart of hipster country and throw a lavish party. But in Dietrich's view it had succeeded—the tone of the lay press had turned positive, and a few days earlier even stodgy old *Nature* had thrown personal genomics companies an editorial bouquet:

> . . . to advocate relatively light regulation does not mean turning a blind eye to the risks of such a strategy. It means taking seriously the presumption that people should be free to inform themselves and make their own choices, and that by doing so they may benefit not just themselves but also the overall pace of innovation.[36]

And no, Dietrich admitted, downtown New York City was not an accidental choice for a launch site; it was market research at work. "In coming to understand our demographic, it was clear that these were forty- to sixty-year-olds, high income, mostly urban, early adopters, fifty-two percent male," Dietrich said. "Where's the epicenter of our customer base? Well, New York is the capital of the world! In SoHo we have our demographic walking down the streets every day."[37]

There was only one problem. Navigenics was forbidden to peddle its product in New York State.

A few days after the SoHo shindig came to an end, the New York State Department of Health sent a hangover in the form of a warning letter to the company and twenty-two others telling them that they needed a permit before they could offer their tests and services.[38] Stephan was unruffled—he told me that it was a matter of submitting applications and paperwork and that he had been to Albany a couple of weeks earlier to meet with Department of Health officials. "Their biggest thing is they want this information to be used in conjunction with a physician. That's totally reasonable. We will probably need to modify the business model a little bit in New York State. We're working through that, but we're good to go everywhere else."[39]

Or so he thought. In June 2008, California followed New York's lead and sent cease-and-desist letters to thirteen companies, including the Big Three (Navigenics, 23andMe, and deCODEme), some of which claimed to be in compliance with state law, while others stopped offering their services in the state, at least temporarily.[40] The California Department of Public Health complained that 1) the clinical labs used by the companies to perform the genotyping were not appropriately certified; and 2) as in New York, the companies could not offer their tests directly to consumers in California without a physician's order. These companies were, according to California officials, "scaring a lot of people to death."[41]

That summer the Secretary's Advisory Committee on Genetics, Health, and Society convened its biannual meeting. SACGHS was formed in 2002 (an outgrowth of an earlier committee on genetic testing) by then-HHS secretary Tommy Thompson, and asked to consider the impact of genetic technologies on American society.[42] The meetings took place in the Hubert H. Humphrey Building, a singularly ugly 1970s edifice that serves as the headquarters of the Department of Health and Human Services and lives in the shadow of the Capitol. Day one of the meeting was as boring and Kafkaesque a gathering as I'd ever attended, devoted to a dissection of the process by which the committee would decide what its priorities were; I thought I'd stumbled into a scene from the film *Brazil*. Two hours into the session my posterior was numb and I could feel my eyes rolling back in my head. Worst of all, the building didn't seem to have wireless Internet available to civilians; I made a mental note to bring an Ethernet cable, some sudoku, or something to read next time.

Day two, unassumingly titled "Session on Personal Genome Services," was the antithesis of the previous day's sleepy proceedings. On three sides of the rectangular assemblage of tables sat committee members—academic physicians, geneticists, biotech executives, law professors, theologians, social scientists, and assorted bureaucrats. At the head of the room were empty chairs reserved for the leading lights of the new wave of personal genomics: Dietrich Stephan (Navigenics), Linda Avey (23andMe), Jeff Gulcher (deCODEme), Ryan Phelan (DNA Direct), and George Church (Personal Genome Project; Knome [about which more later]). There was a cacophony of excited chatter beforehand. By the time the chairman asked people to take their seats it was standing room only.

At this stage of the game, eight months since the launch of the first commercial "Saliva Diviners," the blogosphere had, for the most part, warmed to personal genomics.* Customers were comparing results from different companies;[43] some were writing their own software to interpret it.[44] But the folks in white coats remained unimpressed: the backlash against the companies by the medical establishment was almost immediate. Not only are these companies doing more than dispensing "information," the critics said, they are practicing medicine, and indeed, they are practicing—in the immortal words of Bon Jovi—*bad medicine*. These tests

---

* A notable exception was the Gene Sherpa (http://thegenesherpa.blogspot.com/), run by a clinical genetics fellow, Steve Murphy. Murphy was engaged in his own high-risk entrepreneurial pursuit, the first freestanding genomic medicine practice. From the first he was a vocal critic of personal genomics companies.

are inaccurate, they don't measure what they are supposed to measure, they are not actionable, they will lead to the "wrong" actions, they will lead to unnecessary anxiety, they will rob medicine of valuable resources, physicians will not be able to handle the data deluge, and they are frivolous manifestations of important science.[45] (Perhaps the unkindest cut was "recreational genomics,"[46] a label Navigenics gave to 23andMe and one that, much to Linda Avey's chagrin, had stuck.)[47] When it threw down the gauntlet in January 2008, the *New England Journal of Medicine* encouraged doctors to tell patients that the information derived from these services was essentially useless and that patients should "ask again in a few years."[48]

One of the authors of that editorial was Muin Khoury, director of the National Office of Public Health Genomics at the Centers for Disease Control and Prevention. He was also on the Secretary's Advisory Committee and expected to be a sharp interlocutor at the July meeting. He didn't disappoint.

"Where is the balance between waiting and validation?" he wanted to know. The development of valid cholesterol testing, he pointed out, took many years. "What is the value added by genes versus pure family history and traditional risk factors?" Later, he answered his own questions with a sharp rhetorical flourish: "This information is not ready for prime time. Replication is not clinical validity. When you [panelists] talk about value to consumers and clinical utility, I talk about balance of harms and benefits. The problem is lost in translation."[49]

The companies had also appeared a couple of months earlier at the Cold Spring Harbor meeting. There, at a session moderated by Francis Collins on commercial personal genomics, Dietrich, Linda, and deCODE's Kari Stefansson were downright deferential to the assemblage of scientists. They wanted to know how they could work with scientists. They acted not like competitors, but like colleagues—even deCODE's notoriously cantankerous Stefansson made nice.[50] At the SACGHS meeting two months later, the companies were still "on message": they were doing everything they could to cooperate with the state health departments in New York and California, they were going to meet to hammer out standards, and they were actively soliciting advice from all of the stakeholders in personal genomics.[51]

If there was a panelist who exuded above-the-fray gravitas at the SACGHS gathering (as opposed to Gulcher's slight hostility and George's phlegmatic indifference), it was DNA Direct CEO Ryan Phelan. Like Linda Avey, she was a Silicon Valley veteran. She was married to Whole Earth Catalog author and counterculture/cyberculture icon Stewart Brand; the pair lived on a tiny tugboat in Sausalito.[52] Phelan came to genetics well into her career. But DNA Direct preceded and was wholly distinct from the Big Three and other personal genomics SNP-chip companies; it was in essence a clearinghouse for traditional, one-disease-at-a-time genetic testing. When I met her for lunch in San Francisco not long after the 23andMe and deCODEme launches, Phelan had already sussed out what the new companies would mean for her business. "Both 23andMe and deCODEme are drawing a line and saying, 'We're just health information. We're going to give you some interpretation but not a whole lot. And we're also going to tell you that what we do is not a medical diagnostic. So if you really think you're at risk for something [like breast cancer], you need a follow-on DNA diagnostic.' DNA Direct can be a service to those customers."[53]

Looking ahead, Phelan saw an opportunity in the arrival of whole-genome sequencing: interpretation. "The core competency I want to add is helping people to utilize this cool new technology."[54]

After the regulatory crackdown, the DNA Direct website included a statement from Phelan:

"DNA Direct has not received a cease & desist letter from the California State Department of Health. And DNA Direct has no reason to expect to receive any such letter. Our company fully complies with all applicable state and national regulations for genetic information services, including facilitating genetic test requests."[55] Clearly the company was not in the crosshairs of regulators the way the genome scanning firms were.

Kevin FitzGerald, a SACGHS committee member and perhaps the only person in the world who was both a trained molecular biologist and an ordained Jesuit priest, asked the assembled purveyors of personal genomics, "Would you sacrifice your bottom line for the sake of health care?" Phelan didn't hesitate: "From an investor's perspective, we have *always* sacrificed the bottom line."[56]

Linda Avey chimed in that while 23andMe was a business, it also had a social mission. Jeff Gulcher said deCODEme was genotyping some people at cost. Dietrich seemed piqued by the question. "Implementing genotyping as a service doesn't fit as a not-for-profit model. What portion of medical infrastructure operates as not-for-profit?"[57]

It's doubtful the meeting changed any minds. During a break I heard geneticist Jim Evans, who had earlier compared personal genomics to astrology, say to some attendees, "I don't care if they want to make a buck. They just shouldn't pretend like they're doing something else."[58]

But even if there had been an understanding between personal genomics companies and the feds on that day in July, political reality made it extremely unlikely that anything meaningful would happen before President Bush left office six months hence. Because everyone knew that, much of the day's later proceedings had a relaxed, valedictory feel. Not only did Francis Collins make his last appearance as an ex officio committee member before stepping down as director of the National Human Genome Research Institute in August 2008, but HHS secretary Mike Leavitt himself turned up unannounced (it was *his* committee, after all). And like any capable politician, Leavitt managed to make people feel good while saying absolutely nothing of substance. When asked how he thought we might move ahead with direct-to-consumer genetic testing, he said that he would not weigh in on either side but that the matter represented "a positive struggle." When asked politely by Collins when reimbursement for preventive health-care measures would be implemented, Leavitt admitted that it was indeed a pressing issue. But, he said, "It's not likely to happen in the next one hundred and ninety-seven days."[59]

A month earlier, molecular biologist Steve Brenner of the University of California at Berkeley organized a dinner at a swanky San Francisco hotel aimed at helping to realize his vision of a "Genome Commons." Brenner's idea arose from the need to "specifically address the incredible difficulty of integrating and interpreting multiple variants in an individual, each associated with a given trait."[60] In other words, we will soon have tens of thousands of genomes . . . how in the hell will we even begin to interpret them? As an example of what not to do, he pointed to the 2007 paper announcing the completion of Craig Venter's genome,[61] and especially the "inscrutable" table therein that was meant to relate his genotypes to his traits, but contained only a smattering of random phenotypes and susceptibilities (for example, "evening preference," "novelty seeking," "tobacco addiction," "Alzheimer's"), a sort of 23andMe lite. "What did we learn [about Craig Venter's phenotype from that paper]?" said Brenner. "Nothing that was worth even reporting correctly."[62]

Among others at the dinner were the Big Three: Gulcher from deCODEme, Avey from 23andMe, and Stephan from Navigenics. At that time it seemed that wherever one was, you could find the other two. Geneticist Hugh Rienhoff (whom we will meet again later) opined that most of the variants the companies were reporting to customers didn't pass the "so what" test. A variant that raised one's risk for any given disease from one in a thousand to one and a half in a thousand, for example, was not clinically meaningful. And even when changes in risk were large, physicians still weren't getting the message, said Rienhoff; in fact, they probably didn't care.

"I find every paper that's published in the *New England Journal* about [genetic] association studies to be almost completely useless for the reader," said Rienhoff. "And the [intended] reader is the clinician! But he doesn't understand the statistics, he doesn't understand the technology, and it won't make any difference for ten years."

"But that's what the missing piece is," Avey shot back. "We don't have a way to translate this into the clinic."

"Well," said Rienhoff, "I think it's premature, to be honest with you."

"It's premature," echoed Avey. "But when is it mature?"[63]

This was—and still is—the question for commercial personal genomics, the same one posed by Rabbi Hillel some two thousand years earlier: If not now, when?

Both 23andMe and Navigenics were born of this frustration and impatience. And both companies may have shared a catalyst.

In November 2006, a *Wall Street Journal* article about Augie Nieto, inventor of the Lifecycle exercise bike, described his tragic diagnosis of amyotrophic lateral sclerosis, or Lou Gehrig's disease, at the age of forty-eight.[64] Like breast cancer, ALS is mediated mostly by genes of strong effect in 5 to 10 percent of cases and partially mediated by a collection of much weaker genes in the other 90 to 95 percent. Even had Lou Gehrig never made his extraordinary "luckiest man on the face of this earth" speech, ALS would still be among the most heart-wrenching of diseases. While the body breaks down, the limbs atrophy, and one loses the ability to move and speak, the mind remains intact. It is an agonizing, inexorable process; most ALS patients die within five years due to respiratory failure.[65]

Nieto, however, had financial resources most other ALS patients didn't. He was not going to succumb without a fight. He assembled a team of physicians and scientists, and through his newly hatched ALS foundation, "Augie's Quest," sponsored research into the genetic basis of the disease with the explicit goal of finding a cure. Among the researchers was Dietrich Stephan, then at the Translational Genomics Research Institute in Phoenix. At Nieto's urging and with his financial backing, TGen collected 1,250 ALS samples in three months. The TGen team then typed the samples for hundreds of thousands of markers and assembled a list of twenty-five promising ALS susceptibility genes. Nieto's doctor asked for his patient's genotypes. Stephan wanted to oblige him but found that he could not.[66]

There were two main issues. "The genotypes were done in a research environment and not in a CLIA environment."* In other words, TGen's

---

* CLIA stands for Clinical Laboratory Improvement Amendments of 1988. It is the mechanism by which the Centers for Medicare & Medicaid Services regulate clinical laboratory testing of human specimens. There are some 189,000 CLIA-

lab was not a certified clinical diagnostic lab and therefore the institute was forbidden from returning results to research participants. "Second, the IRB mandates total de-identification of research subjects/samples."[67] The Institutional Review Board (IRB), that is, the ethical review board charged with approving, monitoring, and reviewing TGen's biomedical research on humans (every hospital and academic medical center has one or more IRBs; I serve on Duke's), demanded that research samples be kept anonymous in order to protect subjects' privacy and confidentiality. Whether the subject wanted to waive that right was immaterial. Augie Nieto could not view his own genetic data.

The investors behind 23andMe saw the *Journal* article. "This is why your company has to be here," they told Linda Avey and Anne Wojcicki. "To give people access."[68] (And presumably to fill a market niche.)

Dietrich Stephan conceded that there was another issue, the same one he had confronted when trying to move complex genetic tests into clinical practice. "At the time [2006] we had absolutely no understanding of how to communicate probabilistic risk based on SNPs of very low effect size and what SNPs met the threshold for being 'real.'" In other words, while a particular genetic marker might appear in ALS patients more than it does in controls, perhaps even a lot more, that was just the beginning; there might be ten, twenty, or fifty such markers. And even if there weren't, an association between marker and disease was still a long way from understanding the significance of that marker, what it meant in the context of other DNA markers and/or the environment, and how it might be used to develop therapies. "We still don't know [what's 'real'] for ALS. It's actually one of the difficult problems that prompted the formation of Navigenics."[69]

One could argue that Knome, too, arose from similar circumstances: demand for a service that was not yet available. As soon as news of the

---

certified labs, most in the United States. CLIA is meant to ensure accuracy, timeliness, and reliability of lab test results. Some people—like me—are not convinced it does that, at least with respect to genetic testing.

PGP became public, George Church began to get requests from wealthy people who were prepared to have him sequence their entire genomes on a fee-for-service basis. While he was encouraged by the fact that people were interested, it got to be a headache. "I felt this would be distracting from our academic mission both for my lab and for the PGP, both of which are active in nonprofit research operations. This seemed to me to be a textbook case for starting a company: to get it out of my hair. I thought it would be a good way of calling people's bluffs and making sure they actually *did* want a whole-genome sequence."[70]

Church founded Knome (he pronounces it "Know me"; the CEO says "Nome") in 2007; it began enrolling customers at the end of the year. For $350,000, you could get your entire genome sequenced—all 6 billion base pairs, five to ten times over.[71] When I mentioned it to my frugal wife she raised her eyebrows. "You're getting quite a discount," she said.

What do these origin stories mean? Commercial personal genomics was brought to term via multiple paths. For 23andMe, starting a company was a way to circumvent the inadequacies of publicly funded biomedical research and bring a "holistic view of genomics" to the masses, that is, genetics and self-knowledge in the form of social networking and ancestry. For Navigenics it was a way to make complex medical genetic risk information available to eager (and presumably well-heeled) consumers. For George, starting Knome was simply a way to make commercial personal genomics go away, to segregate it cleanly from his academic and nonprofit enterprises.

None of this is to say that the principals, with the possible exception of George, were not interested in making money—I am not that naïve. But there had to be easier ways to make money. And to lump the top-tier personal genomics companies in with Internet-based vitamin supplement salesmen and other varieties of modern snake oil or late-night infomercial fare was both facile and unfair. This was commerce, yes, but it was also rebellion. And it was dubious commerce, at least initially. While the VC dollars continued to flow, by 2010 no one had gotten rich selling personal genomic services to the public. Indeed, some had already lost their shirts.

In the near term, the elephant in the room would remain determining what all of this information meant. But what was "near term"? When discussing the PGP with friend and Broad Institute geneticist Stacey Gabriel, I said without thinking, "I'm not so worried about interpretation of the sequence. We have the rest of our lives to do that."

"That's good," she said. "Because it's going to take that long."[72]

Stacey's colleague Pardis Sabeti conceded that we were still in the very early days of this stuff, but insisted that that was not the point. "It's unavoidable. This knowledge will be accessible and people will access it."[73]

Statistical geneticist Tara Matise was more agnostic. She had arranged to have her father get his APOE genotype, since Alzheimer's was his biggest concern. As for herself, she could afford it but was in no hurry. "My family is pretty healthy, luckily, and I have not made time to think about how many surprises I want."[74]

I was still many, many months away from getting the full sequence of my twenty thousand genes, but a few weeks before the Navigenics soirée, Jason Bobe sent me an email: " . . . here is your snp data. I've taken a look at it and I'm sorry to report that it's pretty much all junk DNA."[75]

Finally—something to peruse! But jeez Louise, the raw data file went on for days. I would need help to get through it. And I would get it. But even then, of the half million SNPs George's lab typed me for, I would take a hard look at only less than three hundred.

And even that slight peek inside of Pandora's box was more than enough to blind me.

5

Better Living Through Chemistry

Even though it was early February, with temperatures in the sixties and a gusty wind buffeting the Gulf Coast, the lobby of the Marco Island Marriott Beach Resort still smelled like coconut. And not cheap suntan-lotion coconut, mind you, but the actual coconut one would pull off a tree and crack open with a hammer—sweet but not overpowering; an upscale pheromone evocative of drinks with umbrellas, bikinis, white sand, and turquoise surf beckoning from just twenty yards away. Flip-flopped and sunglassed tourists waddled in from the heated pool and assorted Jacuzzis.

Sprinkled among them were a few hundred academic genome scientists and their industrial counterparts—that is, representatives from a select group of companies who cater to university molecular biology types. The nerds and the suits both expected to be pampered, and the opening night reception of the Advances in Genome Biology and Technology meeting suggested that they would not be disappointed. In the dark, spread out in the lush grass next to the pool, there was a tiki motif at work: palm trees, torches stuck in the ground, a band warbling Jimmy Buffett covers in the background, various meats on and off the bone, a bevy of other fresh food stations, and an open bar stocked with beer and wine and rum. The staff-to-attendee ratio was high: a festive-shirt-wearing person was always ready to help, to pour coffee or clear away one's plate.

In part the lavishness came off almost as an act of defiance. With the memory of the 2005 hurricane season and the malevolent sisters Rita and Katrina already receding after a couple of years, this bit of Gulf Coast seemed intent on behaving as though it were oblivious to its precarious location, let alone to whatever curveballs La Niña and climate change might hurl its way. The Marco Island beach was, as ever, dotted with wealthy pensioners moving in and out of their space-age, Bauhaus-on-steroids condos.

At the meeting itself, the opulence was subsidized. When I checked in, the clerk at the front desk handed me two nondescript key cards inside a Marriott envelope. I asked where registration for the genomics meeting was and he stopped short for a moment and then pulled back the envelope before I could pick it up. "I am going to give you different keys, sir," he announced. The new ones were emblazoned with a golden double helix and Applied Biosystems, Inc.'s catchphrase for its newest DNA sequencing machine: "The next generation is SOLiD™." (SOLiD™, as I would be reminded on many occasions over the next three days, stands for "sequencing by oligonucleotide ligation and detection.") I felt as though I had been given a backstage pass to a rock concert. But that was only the beginning: ABI and a handful of other sequencing companies had underwritten the coffee breaks, the meals, the poster sessions, and the boatloads of goodies stuffed inside our complimentary High Sierra backpacks—the pens, the lab notebook, the candy, the beach balls, the digital timer. Plenty of swag for my daughters.

DNA sequencing has only existed since the 1970s and neither of the two original methods was ever patented.[1] So how did it become a multi-billion-dollar bonanza? Until the last couple of years, the overwhelming majority of DNA sequencing was based on a principle developed by the unassuming English biochemist Fred Sanger, earning him the second of his two Nobel Prizes in 1980 (the first, in 1958, was for figuring out how to deduce the sequence of amino acids that make up proteins, the end products whose identities are embedded in the DNA code). Around the same time, a chemically based method of DNA sequencing was developed by George's mentor, Walter Gilbert, and Gilbert's student, Allan Maxam. Both methods were labor-intensive in the beginning; Sanger's method was easier to automate and eventually overtook Maxam and Gilbert's. And even though one could not generate much sequence in a single experiment in the early days, Gilbert said that something had shifted. "In 1975 Allan and I and Fred made sequencing a [laboratory] staple. We changed the problem from impossible to pretty easy."[2]

Sanger's DNA sequencing method, which seems to me no less ingenious now than it did when I first learned about it in the 1980s, exploited the same enzyme our cells use to manufacture DNA: DNA polymerase (most enzymes bear the suffix -*ase*). Essentially, "Sanger sequencing" involves putting DNA polymerase in a tube along with the DNA one wants to sequence, plus DNA building blocks, or deoxynucleotides, each of which contains one of the four DNA bases, adenine (A), thymine (T), guanine (G), and cytosine (C). The "deoxys" are the raw material that the polymerase enzyme uses to extend the DNA chain. But the key to the method was that Sanger also added a small quantity of slightly altered versions of the bases. These "dideoxy" versions could also be added to the growing chain the polymerase is churning out, but each one is a dead end—the chain cannot be extended from a dideoxynucleotide: imagine a section of railroad track with a bumper at one end preventing the track from being elongated. Thus, after the enzyme does its work, the test tube is full of DNA strands of various lengths, each one capped at a random place by a chain-ending dideoxy A, T, G, or C. If one could resolve those chains by size, Sanger reasoned, then it should become possible to read the sequence of a DNA molecule from one end to the other: each nucleotide like the rung of a ladder.[3]

But how to resolve the different-sized molecules? For nearly two decades the method of choice was a gel made of a thin layer of a latex-like chemical called polyacrylamide. The principle is simple: when exposed to an electric field, shorter DNA molecules migrate through a polyacrylamide gel faster than longer ones. Thus, by loading the contents of the test tube full of different-sized DNA fragments into a vertical gel and cranking up the voltage, one could get an overlapping ladder of DNA molecules and read it in order. Automated Sanger sequencing reads yield eight hundred bases per run, sometimes more (the protein-coding portion of a typical human gene is about two thousand bases long).[4]

This was all well and good—exciting even: Sanger sequencing meant that the genomic Rosetta Stone could now be sounded out, even if it wasn't clear what the words meant. For me, reading a clean piece of DNA sequence that might be harboring a disease-causing mutation was one of the thrills of my graduate student experience. But the setup—the "workflow," as the corporate sales reps call it—was decidedly less thrilling. Pouring gels between pairs of glass plates, starting over when they developed bubbles, letting them solidify ("polymerize"), loading them with extreme care, disassembling them several hours later, transferring them to paper and exposing the paper to film, and then reading the sequence by hand . . . all of it got to be a drag. When I was working on my master's in the late 1980s, my adviser used to walk through his human genetics lab and insist to his technicians, students, and postdocs, "This is *not* a factory." Methinks he protested too much: We had an assembly line where we performed nearly identical experiments examining genetic markers, running gels, and seeing what came of them. Day after laborious day, each one divided from the next only by lots of beer and a little bit of sleep. It wasn't quite assembling widgets, but I'd argue it was every bit a factory. And so it became with sequencing and me: I seemed to be most successful when I got into a kind of Zen state and didn't overthink things. (For me, this was difficult—without strong medication of one kind or another, I am a pretty lousy Buddhist.)

Whatever its sweatshop-like qualities, the public and private versions of the Human Genome Project initially used more or less the same assembly-line Sanger approach. Several of the major sequencing centers hired dozens of people whose only job was to pour gels; they would often come in to work in the middle of the night. Variations of this workflow led to the sequence of a composite human genome ahead of schedule and under budget.* Indeed, within a few years automated capillary DNA sequencing, spurred mainly by demand from the HGP, produced yet another revelation. Suddenly there was no gel; it had been replaced by capillary tubes into which the four sequencing reactions were injected. By the late 1990s, there were two commercial capillary platforms. Molecular Dynamics (later Amersham and eventually GE Healthcare) offered the MegaBACE beginning in 1998. In December of that year, Applied BioSystems (then PerkinElmer) began shipping its PRISM 3700, which was eventually succeeded by the 3730.[5] By then the game was afoot: the major taxpayer-funded public sequencing centers had had a fire lit under them by the upstart Craig Venter, the iconoclastic public face of a private initiative that wound up sequencing the human genome in parallel to—and in competition with—the government-funded Human Genome Project. With Venter's heretical commercial entry into human genome sequencing and ambitious plans to annotate and sell the information, there was a huge, ready-made market for whichever sequencing platform could walk the walk.[6]

And what of the face-off between the MegaBACE and the 3700? "It was a real pissing contest for eight or ten months," recalled Steve Lombardi, who managed ABI's sequencing operations in the Americas at the time. "But the 3700 was a better instrument. Better chemistry."[7]

And more savvy marketing. Well before Molecular Dynamics did, ABI saw an opportunity to arm both sides of the fight with its new sequencer. For his part, Venter ordered 230 of the 3700s. On the public side, Eric Lander's genome center at MIT ordered another 115. "By early 2000," Lombardi told me, "the game was over."[8]

---

* Or, as George Church would say, 93 percent of a genome was sequenced. Seven percent remains refractory to sequencing; it is sometimes called "dark matter."

For the next seven years, Sanger sequencing, as embodied by the ABI machine, remained an entrenched technology in hundreds of academic sequencing facilities and biotech companies. Even as late as 2009, Sanger-based DNA sequencing was still very much a growth business—oodles of cash were spent every year on half-million-dollar instruments and the various chemical reagents needed to keep them churning out the endless stream of A's, G's, T's, and C's, the simple digital DNA alphabet that encodes the tens of thousands of proteins that comprise life on earth.[9]

But after the completion of the HGP . . . then what? What would we do with sequencing technology? Now that we'd completed our "moon shot," was there something useful that could be done with the leftover launchpads and rockets? Both Craig Venter and his rival Francis Collins encouraged the technologists to continue to bring down the cost of sequencing dramatically—from the average HGP cost of $1 per finished base (itself down from $10 per base in 1990) to six orders of magnitude less: the $1,000 genome. Of course, this meant that the old rocket fleet would have to be either souped up in a quantum way or else mothballed in favor of new machines.

Accordingly, in 2003 the J. Craig Venter Science Foundation announced that it was offering a onetime $500,000 Genomic Technology Prize intended to goad the research community into advancing automated sequencing to the point where a $1,000 genome was feasible.[10] In 2004, perhaps in part as a response to Venter, the National Human Genome Research Institute awarded $38 million in grants to develop novel sequencing technologies aimed at sequencing a "mammalian-sized" genome for $100,000 as an interim step on the path toward the $1,000 genome.[11]

And in October 2006, the X Prize Foundation announced the launch of the Archon X Prize for Genomics,[12] a $10 million bounty to be awarded to the first private team to sequence one hundred complete human genomes to high accuracy in ten days—a $10,000 genome. At that time, George Church was on the rules committee for the X Prize and had all but ruled out competing for it himself. A year later not only had he decided to throw his hat into the ring using his polony technology, but with a trademark blend of confidence and modesty he predicted that his lab actually had a pretty good shot at winning. But why? What had changed?

"For one thing the rules have sharpened up," he told me. We were driving through the Palo Alto rain after a dinner with several Stanford alums and their Asian backers, entrepreneurs who were starting yet another next-generation sequencing company, this one to be called Light-Speed Genomics. For George, the attraction of the X Prize was the incentive it offered to create an infrastructure able to do large amounts of whole-genome sequencing fast—the infrastructure needed to complete, say, the Personal Genome Project. "It would be amazing to be able to sequence a hundred genomes in ten days, but then what?" said George. "You'll be all dressed up with nowhere to go. Wouldn't it be nice to have a project waiting in the wings that could use that infrastructure and all those machines?"[13]

George had also come to believe that the X Prize shared some of the Personal Genome Project's educational mission, namely, to raise the profile of genomics in the public consciousness, although he thought the X Prize's approach was "a bit more flamboyant" than the PGP's. He may have had misgivings about that, but he was a realist: people liked prizes and competitions. They had a seemingly bottomless appetite for shows like *Dancing with the Stars*, *America's Next Top Model*, and *American Idol*. "I would run into people and talk about how the PGP was going to inform us about genetics and medicine," George said, "and their eyes would glaze over. And then I would mention the X Prize and they would get all excited about it. Like, 'Where can I put my money?' And I thought, 'Gee, this is kind of weird. There are actually people out there who would rather invest in a yacht for the America's Cup than they would in helping to provide for vaccines in Africa.'"[14]

Many of the fruits of each of these various incentive programs to lower sequencing costs, some more ripe than others, were on display at Marco Island. The race to succeed Sanger sequencing was on like Donkey Kong.

"You picked the right year to come," said Ian Goodhead of the Sanger Institute, the British sequencing powerhouse funded by the Wellcome Trust charity. "Last year [2006] it was all 454. By the end of the meeting people were pounding their heads on the table hoping to never hear those three digits again."[15]

As far as I could tell, 454 Life Sciences came into being as a result of genomic determinism. Its founder, scientist Jonathan Rothberg, spent a long dark night in 1999 when his newborn son was rushed to intensive care for an unknown ailment (the boy is healthy today). "Why can't we just sequence his genome and know if everything is fine or not fine?" wondered Rothberg, presumably sure that a genome sequence would indeed be the definitive test. His son's illness notwithstanding, that night spawned an idea. Rothberg had read about Intel's new chip, which incorporated 44 million transistors. Transistors had long since displaced vacuum tubes.* Sanger sequencing technology—the molecular version of the vacuum tube—had hit a wall. The time had come, thought Rothberg, to develop DNA sequencing's answer to the transistor.[16]

The 454 approach was based on three innovations that made it distinct from Sanger sequencing. One was to miniaturize everything: the smaller each DNA reaction is, the fewer chemical reagents it requires, the less space it takes up, and the cheaper it is to run. Rothberg was able to shrink the wells that held each DNA reaction such that four would fit on the end of a single hair.

The second novelty to be commercialized by 454 was to sequence the DNA in "massively parallel" fashion. Typical Sanger sequencing reactions can produce read lengths of eight hundred DNA letters or base pairs. Those are more than long enough to align and use to map back to the twenty-three pairs of human chromosomes that are themselves each composed of tens of millions of base pairs. But the 454 method and

---

* Many electric guitar players, however, myself included, prefer the warm sound of vacuum tubes rather than transistors in our amplifiers. We are both Luddites and fetishists.

several other post-Sanger DNA sequencing methods don't use gels or capillaries to resolve DNA fragments of different sizes and so therefore produce shorter reads.[17] Imagine two jigsaw puzzles of the same dimensions and depicting the same image, one with a thousand pieces, the other with eight thousand. Obviously the second one is going to be more challenging to assemble. For this reason, other investigators had given up on so-called sequencing by synthesis, in which the DNA sequence is read as the A's, G's, T's, and C's are incorporated into the growing molecule instead of being read on a gel or a "trace" of a capillary reaction after the fact (more on that in a minute). In sequencing by synthesis, the sequencing reactions would tend to poop out after only a few dozen bases, yielding a jigsaw puzzle with too many pieces to make sense of. To overcome these short read lengths, 454 went the massively parallel route. As the company showed in the pages of *Nature*, each reaction might produce only one hundred bases, but if you sequenced four hundred thousand wells of DNA at a time, you could still get lots of overlap and lots of sequence; a computer could then do the heavy lifting to assemble the pieces in the correct order and align them to a reference genome.[18] Within a couple of years, the company's sequencer averaged 330 bases.[19]

The other breakthrough of 454 was to use pyrophosphate-based sequencing, or pyrosequencing. Each time a nucleotide (an A, G, T, or C) was incorporated into the growing DNA strand, it would be accompanied by the release of a certain amount of pyrophosphate, a chemical that would provide the energy to stimulate a light-producing reaction. This reaction could be detected and quantified by a camera and easily distinguish among the four bases.[20]

In the early days Rothberg often spoke about 454 technology as a revolutionary phenomenon in a way that presaged the arrival of personal genomics and individual genome scans in 2007. "We democratize sequencing and enable everyone to have a sequencing machine on [his or her] bench," he told *Genetic Engineering News*.[21] "It's completely analogous to personal computers displacing mainframes," he said in *Bio-IT World*. "Now anyone can have their own genome center."[22]

Roche, which had subsidized 454 to the tune of $60 million in the early going, apparently agreed with Rothberg's assessment. In March 2007, Roche bought 454 outright for $140 million.[23]

But as the Sanger's Ian Goodhead suggested when he told me that 2007 was the year to show up at Marco Island, 454 was not destined to be the eight-hundred-pound, eight-hundred-base gorilla that Applied Bio-Systems had been in the sequencing world of the 1990s. Yes, 454 had had a two-year head start on other next-gen sequencing companies and its method was up to one hundred times faster than Sanger sequencing: decoding the complete genome of a bacterium took days on 454's Genome Sequencer 20 as opposed to a month or more with the Sanger-based ABI model 3730. But 454's GS 20 machine had its own limitations. Bacteria were small and they had small genomes: typically a few million bases on a single chromosome. The human genome, on the other hand, was 3.2 *billion* base pairs. Worse, the human genome was a mess, a largely uncharted galaxy where many of the planets couldn't be distinguished from most of the others: some 50 percent of our DNA is known to be repetitive and thus difficult to sequence. In the early going, the 454 machine earned a reputation for struggling with homopolymers—that is, repetitive stretches of the same base. If a piece of human genomic DNA or a tumor sample had, say, a string of eleven A's in a row, the GS 20 might mistakenly call it ten or twelve. That kind of error could really mess up an experiment. The always-tactful Broad Institute sequencing maven Chad Nusbaum initially wondered whether the first iterations of 454 were up to the task of a mammalian genome.[24] And 454 was also expensive vis-à-vis other next-generation sequencing technologies. At Marco Island I spoke to several people from university sequencing facilities who were willing to pay the up-front instrument costs of $500,000 for a GS 20, but worried that they couldn't afford the $100,000 they'd have to lay out for every billion bases they sequenced. Of course the thought of a small university core lab *ever* sequencing a billion bases was itself brand-new; when the public Human Genome Project sequenced its billionth base in the late 1990s, it was cause for a champagne celebration and a PBS camera crew to film the event.[25]

Eventually Rothberg turned his attention to making an affordable bench-top next-gen sequencer a reality. In 2010 he introduced the Ion Torrent machine at Marco Island, a cheap ($50,000) box that could still crank out 150 million base pairs per run.[26]

One of the companies hoping to breathe down 454's neck in the race to supplant Sanger was Helicos BioSciences, a Cambridge, Massachusetts–based start-up tucked away in a *Blade Runner*ish industrial building near Kendall Square, within a few blocks of the Broad Institute. On the wall in the modest reception area was a slogan that was both mission statement and words to live by: "Focus. Teamwork. Safety First. Time Is Short." This summed up the impossible cognitive dissonance that was biotech corporate culture in the early twenty-first century: concentrate on your job and excel at what you do, but be a team player and don't rock the boat, be selfless, be careful, don't rush, don't cut corners. But most of all, hurry the hell up.

Helicos's founder and CEO was Stan Lapidus, a bald, bespectacled engineer and self-described "gizmo guy" with a long history in biotech and molecular diagnostics and, perhaps for that reason, an unflappable air about him (he founded two publicly traded molecular diagnostics companies, Cytyc and Exact Sciences, and endured some bumpy times with both). "I know we will have start-up issues, I know we will have emotion. But I know we will overcome it and things will be fine," he told me after a quarterly call with somewhat skeptical investment analysts.[27]

For Lapidus, the idea to start a next-gen sequencing company came the old-fashioned way: by reading the literature. In a paper in *Proceedings of the National Academy of Sciences*,[28] Stanford's Steve Quake described a method to sequence DNA not from an amalgam of enzymatically amplified DNA, but from a *single* molecule, a then-unprecedented approach. Being able to interrogate DNA *directly* without having to amplify it or otherwise manipulate it was a real milestone, even if the Quake team was able to read only a measly five bases. Lapidus devoured the Quake paper and saw the future. "It was an 'aha' moment," he said. "If luck favors the prepared mind, then I had been loitering in this area for many years."[29]

Most DNA sequencing methods, including Sanger and more recent methods developed by 454 (bought by Roche), Solexa (swallowed by Illumina), ABI (Life Technologies), and George Church's lab (polonies), required an amplification step. To generate enough DNA to detect and read, DNA fragments were amplified enzymatically in a process known as the polymerase chain reaction. PCR takes advantage of the DNA replication process first postulated by Watson and Crick in 1953 and which was really their keenest bit of insight: the structure of DNA had to be such that copies of it could be made easily. And so it is: the double helix unzips, polymerase enzymes attach themselves to the single strands, and the copying process begins.[30] Today we use PCR to generate millions or billions of copies of any bit of DNA we want. To PCR-amplify a piece of DNA, one puts it in a tube with DNA polymerase enzyme (similar to the one that our own cells use to synthesize new DNA), bits of known sequence called primers to get the reaction going, and plenty of A's, G's, T's, and C's to serve as building blocks for the amplification process. The whole reaction is heated to cause the two strands of the DNA to be amplified to come apart and allow the enzyme to attach and do its thing. An hour later, you can have more than a billion copies of what you started with, which makes it much easier to manipulate and analyze.[31]

PCR was an extraordinary innovation in the 1980s; all of a sudden one could do tons of experiments with tiny amounts of starting DNA. The tedium of cloning DNA into bacteria in order to amplify it could often be avoided. The FBI began using PCR to analyze DNA from extremely minute samples of blood, hair, skin, and semen obtained at crime scenes (remember the O.J. murder trial?). PCR's inventor, the notoriously wacky surfer dude and HIV-denialist Kary Mullis, received the Nobel Prize in fairly short order.[32]

But for all of its advantages, PCR was not necessarily ideal for large-scale sequencing—that is, for sequencing many samples and whole genomes. Rather, PCR was sometimes viewed as just another opportunity to introduce additional expense, manual labor, and human error. And even the process itself could be unpredictable. "Some DNA fragments just don't PCR well, just as some pieces of DNA simply don't clone well," said former Church lab postdoc Jay Shendure, now at the University of Washington in Seattle. Although he was one of the architects behind polony sequencing, the Church lab's PCR-based sequencing method, Shendure readily conceded that as far as large-scale sequencing goes, PCR could be "a pain in the ass."[33]

Stan Lapidus had arrived at the same conclusion independently. In the 1990s he had developed an early genetic screening test for colorectal cancer and had gotten swept up in the work of Johns Hopkins cancer genetics pioneer Bert Vogelstein. After reading through hundreds of Vogelstein's papers, he began to wonder why cancer researchers didn't do what seemed to be the obvious experiment. Cancer arises, we know, because cellular genomes become unstable and this leads to uncontrolled cell division. "If we believe cancer is a disease of altered DNA, then the real experiment is to sequence one thousand, ten thousand, or one hundred thousand tumors," Lapidus said. "So why would a smart guy like Vogelstein not do this simple experiment? The answer was because he couldn't."[34] It was too expensive and too time-consuming, thanks in part to the necessity of having to set up and run so many PCR reactions. Lapidus had long wondered if PCR couldn't be circumvented altogether and thereby simplify the laborious workflow demanded by large-scale sequencing.

"I knew it was a tractable problem," he told me. "I couldn't tract it myself, but I kept my eyes open."[35] After reading Quake's paper, he and the Broad Institute's Eric Lander flew out to Stanford to visit Quake and license his technology. Quake grilled Lapidus, and Lapidus followed up by sending him a mathematical model of a problem Quake had been working on. "We spent about six weeks falling in love," said Lapidus.[36] In another six weeks he had raised $27 million.[37] Lapidus then hired a raft of first-call chemists, engineers, and executives.

Among them was Tim Harris, an analytical chemist by training who cut his teeth at Bell Labs, where, he told me, "AT&T spent a hundred years alienating the engineers and scientists." Wanting a new challenge, Harris moved from a field where people were trying to make new materials to one where that step was already done. "In biology, God's already made all the stuff." He joined a start-up called SEQ; his team developed an auto-mated imager ("a nice scope," he called it) that allowed drug developers to examine the effects on cells of thousands of compounds at once—a real time-saver. The success of the imager led to SEQ being sold to British life-sciences giant Amersham (now part of GE Healthcare), but Harris felt his new bosses never really "got" what it was they had bought. "We tried to teach Amersham how to make this thing we'd invented, but were met with limited success. I've found that technology doesn't transfer, only people do. What was broken [at Amersham] I couldn't fix."

He emailed Stan Lapidus, whom he'd met years earlier and who shared his interest in biological measurements, and said he was leaving Amersham. Lapidus quickly wrote back and asked when he could come to Boston. Harris was one of Helicos's first two hands-on, tech-oriented employees. "Single-molecule sequencing was hard. I thought there was a fifty percent chance we would fall on our faces," he said of the company circa 2004. "Maybe Stan didn't."[38]

Lapidus assembled a scientific advisory board of thought leaders in the field, including Quake, George Church, and automated sequencing pio-neer Leroy Hood.[39] Helicos's board of directors came to include venture capitalist Noubar Afeyan, the guy who launched Craig Venter's commer-cial sequencing venture Celera Genomics while still at ABI.[40]

It was an impressive lineup. But all of this after reading a paper that outlined a method that had managed to sequence only five bases at once? The average gene was two thousand bases long. The human genome was 6 *billion*. What exactly was Lapidus thinking?

"Quake had addressed the fundamental physical and chemical ques-tions behind single-molecule sequencing," he said. "The judgment you make in the entrepreneur game is this: When is enough proof-of-principle enough? I've had three at-bats in my career, and in terms of assessing technical feasibility, I have a pretty good track record. That may be the one skill I have."[41]

Lapidus contacted Steve Lombardi, the former ABI executive and battle-scarred veteran of the Genome Wars who'd persuaded Craig Venter to join forces with the company and guided its North American sequencing business from 1989 to 1998. Lombardi, a large and friendly man with a beard, glasses, and New England roots, was trained as a nucleic-acid chemist. After seventeen years at ABI, which was not exactly well-known for its forward-thinking corporate culture, Lombardi decided he needed something else. He moved to one of the other Silicon Valley–based genome-technology behemoths, Affymetrix, and while there worked in corporate development, R&D, and marketing.[42] Affymetrix was a darling of the biotech market in the 1990s as it led the way in the development of microarrays, thumbnail-sized glass chips that could hold tens of thousands of DNA fragments. By probing the chips with a DNA sample of interest (say, from a patient's tumor), one could measure the extent to which an entire genome's worth of genes were turned on or off in any given cell type.[43] Affymetrix went on to develop chips that could allow one to genotype a million or more DNA markers at once, thereby providing the technology used by personal genomics companies such as Navigenics to offer glimpses of customers' genomes (see chapter 4). But Affymetrix had struggled in the new millennium: the bursting of the biotech bubble hit it hard while other companies, most notably Illumina, caught up in microarrays and branched out into other technologies, namely sequencing.[44] Affy was accused of preferring litigation to competition in the marketplace.[45] In the course of researching this book, Lombardi was one of many former Affyites I talked to who had moved on.

But Lombardi was not committed to jumping ship when he went out to dinner with Stan Lapidus in early 2006. It would be a huge professional risk: after all, ABI had ruled the DNA sequencing roost for nearly a decade, 454 had already launched its next-gen machine, and Solexa and ABI itself were poised to join the fray with their own would-be successors to the Sanger method. Whatever Affy's problems, why would

Lombardi leave a comfortable gig there to move back across the country and join a company built on an unproven technology that would be, at best, fourth to market? But he liked the people he saw at Helicos, many of whom he knew personally from his years at ABI. And like Lapidus, he was seduced by the simplicity of single-molecule sequencing.

"I remember coming back from dinner with Stan and my wife said, 'You look like you're thinking about this.' And I said, 'If it works it could be the next big thing.' I was fifty-one, the house was paid for in Palo Alto, and my daughter was a senior in college. If there was ever a time to take a flyer, it was now." The Lombardis packed up their belongings and their 175-pound mastiff, Jezebel, and drove east.[46]

When I asked Steve what was so intriguing about Helicos's "true single-molecule sequencing," his answer was a hybrid of business and science, the take-home message of which was "less is more." Success in the next-gen market was going to be about how much DNA one could pack on the surface of a flow cell—that is, the small surface where the actual business of sequencing took place inside the machine. "We think we're going to be able to more effectively play the game of price versus performance," Lombardi said. "We think this will end up being about the length of read times the number of DNA strands one can fit on a flow cell. And inherently, we are thousands of times better on the number of strands because we can pack single DNA molecules together very tightly."[47]

The other appealing bit of simplicity for Lombardi was the way in which Helicos streamlined the entire sequencing process, a perpetual thorn in gene jockeys' sides—recall Jay Shendure's description of PCR as a pain in the ass. "How did the ABI 3700 transform the sequencing industry in the 1990s?" said Lombardi. "Workflow. With automation, you needed to do less work both before and after the actual sequencing. The same is true for the HeliScope," he said, flashing me a slide of the company's large fridge-sized sequencer. "Here's our sample prep process: purify the DNA from the sample, shear it into small pieces, do a simple enzymatic step, and then load the flow cell. It's a simple process, a scalable process, and a repeatable process."[48]

The next-gen landscape roiled and consolidated: 454 was bought by Roche for $140 million.[49] Illumina bought Solexa for $600 million and quickly became the front-runner in next-gen sequencing, and ABI bought Agencourt Personal Genomics for $120 million.[50] Helicos, on the other hand, opted to walk alone, going public in May 2007.[51] As the other guys went looking for suitors, why did Helicos choose to remain a pure-play, stand-alone sequencing company, especially since it was relatively late to the party?

"It's pretty simple," said Stan Lapidus in a voice that had clearly worked its magic on venture capitalists and investment bankers. "If you believe that this class of technologies changes the way scientists do research, changes the speed with which pharma companies make their way down the funnel to get compounds that might actually work, and changes the way diagnostics is conducted—and I believe all of that as much as I believe anything—then you have to believe that we're not building a company that is to be opportunistically sold. Rather, we're building a company for the long run—we're building a company with legs."[52]

Lapidus went on to say that Helicos's remaining independent made sense because it was not a one-trick pony; in his view, single-molecule measurements represented a paradigm shift that extended well beyond genomics. "Until now biological measurement has been carried out from a chemist's viewpoint, measuring stuff indirectly in moles—a wacky way to view things. That makes no sense to a cell biologist because cell biology is based on low concentrations of stuff in a cell overall, but with very high *local* concentrations. In other words, biology is inherently about single molecules. Indirect measurements in biology have slowed progress down, just as they have in every other science. The War on Cancer hasn't moved as fast as we had hoped because we measure cancer indirectly. Another way to think of Helicos is as a company whose mission is to *change measurement science* in biology. Today we're focused on nucleic acids; in the future we may focus on proteins, carbohydrates, antibodies, all on a single-molecule scale. In our view the HeliScope is the universal biological measurement platform."[53]

Investment analysts I spoke with saw more practical reasons for Helicos to go public: money. When Illumina ponied up $600 million for Solexa in late 2006 during a terrifically soft biotech market, it was a clear signal to Wall Street that next-gen sequencing was hot. John Sullivan, a pompadoured Bostonian who covered the sequencing market for Leerink-Swann, said the richness of the Solexa deal "made it easier for Helicos to tell its story to investors."[54] Un Kwon-Casado covered Helicos for Pacific Growth Equities, which, like Leerink Swann, was a Helicos backer. Over lunch on a stereotypically foggy San Francisco day in December, this tiny, amiable, astute woman patiently explained why the company's $43 million net was not all that important. "It's not about how much money you raise," she said, "it's about what your company is worth *after* you raise the money. Today [December 2007], Helicos is worth about $200 million. At this stage they couldn't get $200 million from a pharmaceutical company—it's too early. Someday they could be worth $1 billion. Or they could be worth zero."[55]

At Marco Island, Helicos's chief science officer, Bill Efcavitch (another former ABI engineer), gave a talk that aimed to show how the company had conquered the homopolymer problem—that is, the inability to get the enzyme to correctly read through repetitive regions of DNA. Using a newly invented molecule to make sure the sequencing reaction occurred with high fidelity, the Helicos brain trust presented data on successful sequencing of homopolymers. After the talk, Steve Lombardi came bursting through the doors, buoyant. "People were questioning whether you could do this on a single-molecule platform. And these guys just nailed it. It sort of closes the book on the technology component. Now it's just development at this point, which has its own risks and pitfalls."[56]

Indeed. Helicos burned through $33 million in 2007. At the time of the public offering in May, the expectation was that the first instrument would be shipped by the end of that year. By summer the company had multiple prototypes on its factory floor in Cambridge. But Helicos had been promising to deliver an instrument for years and had yet to do so. Would the company be accused of crying wolf? Not only that, now

Lapidus, Lombardi, and company were constrained by the rules of disclosure and discretion Helicos had to follow once it became publicly traded. Would its silence be interpreted as bad news?

Lombardi assured me that the beta testing was going well, but the machines were still not quite up to the specs the company had promised: 90 million bases per hour at the first pass and 25 million at the second, output that was unthinkable two years earlier but that now seemed necessary if Helicos were to continue to turn heads on Wall Street and in the labs that were considering the other new machines from 454, Illumina, and ABI—machines, incidentally, that were already on the market and whose sticker price was less than half of Helicos's $1.35 million.

"The HeliScope is a pretty expensive machine to make," said Tim Harris. "But $1 million is a serious barrier to potential customers and I tried very hard to convince management of that. I thought they should make a machine that went a third as fast for a third as much. I lost the argument."[57]

"Gosh, the price of that machine would have to get cut in half before I'd even look at it," said a wide-eyed Kevin Shianna over drinks at a Marco Island cabana in 2008. Kevin was a friend who ran the primary sequencing facility at Duke. "Even if that is the best technology out there, imagine us trying to get that money from our funders. Five hundred thousand is a lot of money, but when you start talking about one million dollars just for the instrument . . ." His voice trailed off. "And then you have to deal with the enormous amount of data that comes off of it. Now you have to buy server space. It starts to add up."[58]

Kwon-Casado thought this reaction would be typical. "Seventy percent of the market will not buy this instrument because it's too expensive," she said. "A lot of labs have barely scraped together the budget to buy an Illumina. Helicos doesn't have to be Illumina and move a hundred and fifty instruments a year to be a success. But," she went on, "they haven't launched on time, they haven't published any data, and they haven't shown how quickly they can ramp up production and improve their specs. They have a lot of work to do."[59]

She confessed to being underwhelmed at the company's IPO presentation. "They had this big-picture, fluffy investor presentation talking about diagnostics and all these huge market opportunities. They compared the HeliScope to the microscope—they said that this machine would be to human health what the microscope was to infectious disease! Come on. Just give us the technical details: How are you better than Solexa? How fast are you gonna get there? We were being asked to have blind faith in management that they weren't just bullshitting us."[60]

Meanwhile, just as Helicos was having its Wall Street coming-out party, third-generation sequencing companies were already raising money and generating buzz about technologies rumored to be still faster, cheaper, and more accurate than the current state of the art: Pacific Biosciences, VisiGen, Complete Genomics, and Intelligent Bio-Systems were all starting to court investors and high-profile scientists.

For Helicos, questions abounded: How could the company charge twice what the competition did without demonstrating tremendous accuracy? Where were the data? It was not clear when the first instrument would ship. Folks were getting nervous.

The Broad Institute's Chad Nusbaum was more forgiving. "I understand their desire not to get a machine out before they're ready. They've been very careful over the last year, which I think is good because they misjudged how far along they were a couple years ago. They learned from that experience."[61]

And despite her own stated misgivings, Kwon-Casado, too, was cautiously optimistic Helicos would get its act together. If the instrument did what the company said it could do, she told me, the rest would take care of itself. "Listen, it's a different technology. It's not really competing with the Illuminas and ABIs of the world. It's a different market segment. It's a Ferrari to the everyman's Chevrolet. The Ferrari market is a lot smaller; there's not as much demand. But it has its place in the world."[62]

Ferrari or no, it remained to be seen whether Helicos could traverse the "Valley of Death" that lay between Steve Quake's brilliant invention and a fleet of machines happily cranking out gigabases of DNA sequence.

. . .

Aravinda Chakravarti, my former Ph.D. thesis adviser and an enthusiastic user of the latest and greatest technologies in his human genetics research, was the keynote speaker at Marco Island. We stood in the courtyard sipping beers and absorbing the steel-drum music and that evening's ersatz Jimmy Buffett. I asked him about next-gen sequencing. "It's an interesting time," he said. "Everyone is gorging themselves, but we don't know yet whether it's on caviar or on Cheetos."[63]

Back in Boston, George Church and his merry band of students, engineers, computer scientists, and biologists had another idea altogether: they were convinced they could deliver caviar at a Cheetos price point.

6

And Then There Were Ten

Esther Dyson (aka PGP Subject #3) emailed—all small letters, no punctuation, signs her name "esthr"—to ask if I could come early. But when I showed up to the Washington conference room of the National Endowment for Democracy, where she'd decamped for the moment, another guy was already there to see her. A few months earlier she'd sent me her travel schedule: Moscow, Brussels, London, New York, Aspen, Washington—and that was just June. No wonder she had no land telephone line—it would atrophy from disuse. As a consultant to the air taxi business, she was ferried all over the globe on someone else's nickel, which made her life seem like some kind of *Condé Nast Travel/Wired* mashup of "Where's Waldo?" If I wanted to know where she was, she suggested I follow her on Dopplr, whatever that was.

She had offered advice to countless Web start-ups, to social networking sites, to search engines, to marketers and branders, and to the occasional prime minister. I went on her Flickr stream (why so much disdain for the lowercase *e*?) and looked at photo after photo of exotic locales and famous people from the virtual land of Web 2.0 Capitalism: Peter Gabriel, Sergey Brin, Rupert Murdoch, Thabo Mbeki, Stewart Brand, Neal Stephenson, and Esther herself lounging with some wild cheetahs in Botswana. She hosted an annual gathering for those interested in the emerging markets for private air and commercial space travel. And she was chosen to be the lone outsider to sit on the board of directors of 23andMe.[1] Two decades after the dawn of the Internet, four-feet, eleven-inch Esther Dyson was still the Doyenne of the Digerati, and increasingly, the Genomerati.

I tried to imagine what she was like as a child, with a world-famous physicist for a dad (Freeman Dyson) and an eminent mathematician (Verena Huber) for a mom. A few times a year, dozens of organizations would have their orderly universe disrupted by fifty-something Esther, ever the iconoclast, in a black short-cut jacket, scruffy gym shoes, her hair pulled back and streaked with garish colors, stickers covering her laptop commemorating all of the far-off places she'd been despite never having learned to drive.

I waited for her in the library in the office building on F Street N.W. I had no idea what the National Endowment for Democracy was, but I was pretty sure it was doing something important and paradigm-shifting or else Esther wouldn't be bothered. NED's mission, it turned out, was to foster democratic institutions all over the world.[2] Esther was particularly drawn to Eastern Europe and its burgeoning post–Cold War markets. "It's like a nest," she said. "In destruction lies opportunity."[3] NED was populated by smiling faces, but the place made the Library of Congress seem like a dance party; it was downright funereal, which made Esther stand out. And maybe that was the point: she didn't care about their rules or their culture—she was as honest with them as she was with herself. Therein lay her "value proposition."

In a profile of her from 1996, when the World Wide Web was neither worldly nor wide, she said: "There's going to be so much more content out there, some of it really crappy, some of it not. It's going to be a lot harder to get people's attention, and there will no longer be a premium on

[the] distribution mechanism, which was based on the shortage of channels. Suddenly there are millions and millions of channels. . . ."[4] Had they
only asked her, the record companies, the newspapers, and the television
networks might have saved themselves a lot of heartache. And indeed,
perhaps the "millions of channels carrying dubious content" model was
applicable to human genomes as well.

But even if that weren't true, there would be nothing in Esther's genome that would give her pause. Just like Halamka. And George. And
Stan Lapidus. These people woke up every morning with their swords
drawn, ready to charge up the hill, indefatigable, relentless. In Esther's
case that might mean waking up for an early-morning swim in Kuala
Lumpur or Bombay or Monte Carlo or Frankfurt or Sydney. I envied her
her surefooted place atop Maslow's pyramid. I took comfort in knowing
that when we discovered there were things written in our cells that we
didn't actually want to know, or this whole genome thing turned out to
be a mirage, then people like Esther would be there to absorb the blows.
Scientists and ethicists might have dismissed the notion of hypereducated,
rich, healthy people getting their genomes sequenced first, but part of me
was relieved at the prospect.

Well over a year before we got any actual sequence data, Esther was
appearing on *Charlie Rose*,[5] blogging for the Huffington Post,[6] and writing op-eds in the *Wall Street Journal* explaining the PGP, heralding the
new genetics, scoffing at those who feared it, and radiating sanguinity.[7]
Decoding and publishing genomes, and tying that information into the
medical system, seemed to her a natural extension of the digital zeitgeist.

"There are two big places where information technology should have
helped and hasn't," she told me. "Health care and education." She recalled
a Renaissance Weekend [an intellectual retreat made famous by the Clin-
tons] session she attended with various luminaries in health care, from
insurance company executives to celebrity surgeons. They presented their
uninspired and well-rehearsed shticks to a smattering of attendees. Their
lack of an action plan gnawed at Esther. "They were caring people, they
were lucid about the problems, they were well meaning. But it was the

most depressing meeting I've ever been to. In my world you would have had four VCs and five schemes to revolutionize health care and resolve the obstacles."[8]

She kept getting more and more interested in the subject, which led her to George Church. "I thought, 'Boy, this guy's really interesting but I don't know what to do with him.'"[9] She organized a conference on personal health records, invited him to speak, and then volunteered for the PGP; she soon became Participant #3.

Before the end of the second PGP-10 gathering, Esther had to leave for Houston. She was scheduled for a colonoscopy, part of the medical exam for aspiring astronauts. From there she would go to Russia to complete her training.[10]

In the summer of 2006, the Church-Wu family drove west from Boston to the Great Lakes. Their tour took them through Buffalo, Toronto, Ohio, Michigan, and back through New Jersey and New York. Instigated by Ting and Marie, the journey was an opportunity to meet some of the then-150 people who had put their names in the queue to join the PGP-10 (within two years, the waiting list would reach 5,000; by late 2009, more than 12,000). Even in the early going, George had fielded dozens of phone calls from strangers, many of whom had little if any understanding of the science. "There had been a lot of discussion about who was going to volunteer and why," Ting said. "We thought maybe it'd be nice to meet them and talk to them. We decided the trip would not be to gather DNA but to gather opinions."[11]

"Up until then," fifteen-year-old Marie told me, "all I heard were scientists talking about it in these terms that I couldn't quite understand. I got a lot of it, but it never really seemed that . . . personal? And then I met random people in little towns who understood it more like I understood it. And they had the most interesting genes! It seemed a lot more real to me than just my dad talking about getting his genome done."[12]

Among the random people with interesting genes was Kirk Maxey, a trained physician who founded and runs Cayman Chemical, a leading research supplier of enzymes, antibodies, and the like. He is five feet ten, 155 pounds, blue eyes, right-handed, and his ABO blood type is A+ (PGP profiles come in handy).[13] His self-published book, written under the pseudonym "Sandman," included short stories and essays set in Wyoming, Utah, and New Orleans.[14] But to listen to him speak was to hear a gentle voice whose accent betrayed its Michigan roots, which were quite deep. As was his legacy.

Kirk Maxey may have fathered as many as four hundred biological children.

When he was a medical student in the late 1970s and early '80s, Maxey's wife thought he would make an excellent sperm donor. And so he became a regular.[15] He described his motivations as "50 percent altruism, 25 percent monetary, and 25 percent [something else]."[16]

In those days, the transaction was a casual one. "At that time the OB-GYN or the fertility practitioner just recruited his tennis partner or his cardiology friend from across the hall," Maxey remembered. "He personally vouched for their general healthiness simply by 'inspecting' them. There was really almost no medical aspect to being a donor in those days, just a twenty-dollar bill and a little brown box for your jar."[17]

When he started donating at the University of Michigan, Maxey signed a one-paragraph document whose main purpose was to obligate the donor to report any new sexually transmitted diseases he contracted. "Basically I promised that if something turned blue and fell off then I would tell them."[18]

The form also guaranteed anonymity. In his youthful naïveté, Maxey thought this meant the sperm bank wouldn't tell his med school buddies what he was up to. In fact, it meant that he would remain unidentified to the couples who availed themselves of his sperm, and they in turn would remain unknown to him.[19]

Maxey admitted that he took a lot of things on faith. The first test-tube baby, Louise Brown, had been born in 1978,[20] and in the 1980s in vitro fertilization was a nascent growth industry.[21] Maxey was told that in addition to being used to help infertile couples, his sperm would also be instrumental in perfecting IVF technology. But he never asked how much of his donations actually went toward research.

Like a boxer before a fight, Maxey would have to abstain from sex prior to his donations. But like a porn star, he also had to produce on demand. While the sperm bank would often call two days prior to the date it needed his services, there were times when he would be summoned in the middle of his gross anatomy laboratory—a recipient was ovulating *right now.* Thus he once had to produce a sample while driving. Although he's told the story many times, with a faint smile Maxey described the process of masturbating into a cup while behind the wheel of a moving automobile as "risky and fairly difficult to do."[22] And you thought texting was dangerous.

As time passed, the sperm banking industry became somewhat less free and easy, particularly in the aftermath of the AIDS epidemic.[23] During this time there was also a technical breakthrough: the ability to freeze sperm for later use.[24] For Maxey the urgent phone calls stopped.

But his career as a donor continued, at least until what he called a "really unpleasant and still kind of undefined event."[25] Despite its cold, clinical, anonymous protocols and dirty magazines in the bathroom, sperm donation is an intimate undertaking. Donors often get to know the technicians, mostly female nurses, quite well. With HIV testing came even more intimacy in the form of regular urethral swabs. One of the technicians with whom Maxey dealt regularly had, he suspected, developed a crush on him. He did not give it much thought until she called him one day and told him she had used one of his samples to impregnate herself.

Maxey was shocked at this breach of protocol and what it said about how the sperm bank was doing business. He realized that the clinic had no reliable way of tracking its samples. The physician in charge tried to reassure Maxey that the amorous technician had been fired. "And he said something I always found curious," Maxey told me. "He said, 'We made sure she had an abortion.' I don't see how they did that, though I didn't question it at the time. I didn't want to talk about it. To this day I don't know if she was ever really pregnant or if she eventually had a child."[26]

For a long time Maxey tried to forget his fourteen-year tenure as a sperm donor. Eventually he learned about Wendy Kramer, whose son Ryan was donor-conceived (Wendy was divorced within months after Ryan was born). From age two, Ryan was curious about his biological father.[27] Later, he wondered about whether he had any biological half siblings.[28] Consequently mother and son developed the Donor Sibling Registry, which allowed donor-conceived children to search for relatives by sperm bank name and donor number.[29] At age fifteen Ryan used a combination of a Y-chromosome DNA test and some clever genealogical searching (a relative of his biological father had taken the same test and posted a large genealogy online) to identify his "anonymous" donor father.[30]

"After registering on the site," said Maxey, "it all sort of came back. I said to myself, 'You know, that was a mess—what were those guys really doing?'" With the help of an attorney, Maxey procured seven years' worth of his records from the clinic.[31] These were the source of more revelations.

He found out that his donations were split, sometimes by as much as 8:1. The benign explanation was that this was a way to make donations go further, to share the wealth, as it were. But the real rationale, said Maxey, was more sinister. The samples were being diluted to the point that they were equivalent to an infertile ejaculate. By diluting a sample of 400 million spermatozoa into eight or ten samples, the odds were less than fifty-fifty that a woman would get pregnant. On the other hand, it clearly raised the odds that she would spend another five hundred dollars for an additional vial of sperm, and perhaps another one after that. Maxey believes this is still standard practice in the sperm banking industry. "The most dissatisfied customer is actually the ideal customer."[32]

What about fertility research? Maxey found out that none of his samples went for this purpose; all were used for insemination.

Then came the math. At 100 percent fertility, the clinic's records suggested, Maxey would have produced about 2,600 offspring. Factoring in the dilutions and the randomness of successful fertilization, he suspected the number fell by a factor of ten. "But it's still hundreds," he admitted, a bit incredulous even after two decades. What disturbed him more was that the clinic had no idea as to how many children were conceived with his sperm, where they lived, and whether they were healthy.[33]

Which brings us to his unbridled enthusiasm for making his genome public. He wanted his biological children to be able to contact not only each other but him. They could, if they liked, see what he had in his cells and what they may have inherited without having to resort to any of the elaborate detective work undertaken by the teenage Ryan Kramer. To that end he established the Cayman Biomedical Research Institute ("CaBRI") and its Donor Gamete Archive, which Maxey called first and foremost "a nonprofit storage service."[34] For a fee CaBRI would store 1) viable semen samples that had been purchased by women for artificial insemination; 2) the empty vials that presumably still contained enough residual sample for genetic testing; and 3) cheek swabs of women who had undergone or would undergo AI as well as swabs of their relatives.[35]

CaBRI uses the samples for genetic testing and would test for whatever traits a sample's "rightful owner" (the woman who purchased the sperm sample) requested. It would not try to identify the donor specifically, but the institute pools height, weight, ethnicity, and whatever other information has been obtained about a specific donor and shares it with mothers who have used that donor.[36]

For Maxey, CaBRI was all about trying to reform a sperm banking industry that, at least until recently,[37] appeared to answer to no one. The American Association of Tissue Banks, for example, issued voluntary guidelines that encourage sample traceability; in November 2009 a search found eight sperm banks that were AATB-accredited for both storage and distribution.[38] Much like human genetic and genomic research, the sperm banking industry had spent decades with donor anonymity at its ethical foundation. By tradition the donor is not known to the recipient or the recipient's children. After Ryan Kramer found his biological father, California Cryobank, the largest sperm bank in the world, immediately removed a host of potentially identifying information from its donor profiles in order to make it harder for future Ryan Kramers.[39] And some bioethicists complained that what Maxey's Donor Gamete Archive was doing was not kosher. "Surreptitiously keeping samples in a biobank without explicit consent is unethical," the University of Pennsylvania's Arthur Caplan told *New Scientist*.[40]

But the sperm banks' main argument for preserving anonymity was that without it, no one would donate and sperm banking as we know it would cease to be. They pointed to the United Kingdom, where the Human Fertilisation & Embryology Authority collected data on donor-conceived births, limited use of a single donated sample to a maximum of ten families, and kept a registry of donor data that offspring could access at age eighteen or beyond. Reportedly there were now less than two hundred British men willing to bank their sperm, while Sweden's sperm banking industry suffered a similar decline.[41] On the other hand, this change had served to internationalize the donor conception business. Following their countries' rejection of donor anonymity, both English and Swedish women began traveling to neighboring Denmark on "fertility vacations."[42] And Copenhagen might be just the first stop. "In India there are lots of fertility clinics willing to do any IVF procedure you want, no questions asked," Kirk Maxey told me. "So you order your semen in Denmark, have it sent to India, see the Taj Mahal, and if you have no functioning ovaries yourself, you secure a donor egg and come back pregnant with a half-Danish-half-Indian baby. I know an Englishwoman who did exactly that."[43]

Sperm banks have also had to contend with litigation, some of it resulting directly from their absolutist positions on anonymity and failure to secure and/or track complete donor medical histories. In 1988, a Santa Barbara couple, the Johnsons, used a California Cryobank sample from "Donor 276" to conceive their daughter Brittany, who at age six developed autosomal dominant polycystic kidney disease, a genetic disorder that usually appears later in life but typically requires a kidney transplant by age fifty. The Johnsons had no family history of ADPKD, but Donor 276 did: his grandmother, aunt, and mother all suffered from it. In all, nearly 1,500 vials of Donor 276's sperm were sold by California Cryo-bank to an unknown number of women in unknown locations before the sperm bank pulled the sample in 1991. The Johnsons sued California Cryobank for professional negligence, fraud, and breach of contract, al-leging that California Cryobank knew that Donor 276 was at high risk for ADPKD; the case was settled out of court in 2003.[44]

Recently the sperm banks have shown a willingness to give a little on the anonymity issue, which I suspect stems from the developments I've discussed: the trend toward outsourcing, the ingenuity of and technology available to the Ryan Kramers of the world, the emergence of the Donor Sibling Registry and the Donor Gamete Archive, and the ongoing threat of costly litigation. In 2008, California Cryobank teamed with two other large sperm banks to propose the first national registry of sperm and egg donors.[45] But by 2010, the details—name or just donor ID number? Access for whom and at what cost?—had yet to be ironed out. I had my doubts as to whether they ever would be.

I don't have any data to back it up, but I don't think it's much of a stretch to say that Maxey was not a typical sperm donor. He had and has strong proprietary feelings for his biological children: "[T]hey're my kids," he told PBS in 2006. "They're kids that I adopted out when they were single cells."[46] Thus the Donor Gamete Archive is partly self-serving, and I mean that in the best way. And his book, *Pig Blood*, was written not to achieve literary stardom, but as another way to reach the hundreds of offspring carrying Maxey chromosomes, most of whom were believed to reside within 150 miles of Ann Arbor, Michigan. *Pig Blood* recounts his time working on ranches and in laboratories, and includes pointed essays about evolution (it's still going on) and climate change (it's happening but we'll be fine). The back cover says:

> One day in the spring of 2006, he [Maxey] received an email
> from a young woman in a nearby town who was just about
> to graduate from high school. It said, "I think you are my
> biological father." . . . As this unusual friendship developed,

he learned that they both spoke French, enjoyed writing, and shared a weakness for a certain soft drink. . . . [When her dozens of half siblings came to know the circumstances of their conception], they would come searching for their genetic roots. They would want to know something about the man who helped to give them life.[47]

By the end of 2009, Maxey had found just two biological offspring.[48] "That means that I will die before contacting more than 90 percent of them. That was the reason I wrote my book: so that I could take possession of my own identity, rather than have Google, Fox News and various websites do it for me."[49]

He said that the common perception of the sperm donor is an ugly one: "A testosterone-sodden college kid with his hands perpetually down around his groin, ejaculating in paper cups and then heading home with a couple of six-packs to share his fortunes with his roommates."[50]

He believed that this stereotype could and should be reversed and that genomics would be instrumental in doing so. In the near future, potential donors would undergo full genome sequencing. Those carrying deadly single-gene disorders would be culled from the donor pool much as the Red Cross selects out HIV-positive individuals. The elimination of those without a "Clean Genome" would shrink the donor pool by half. Of those remaining, 90 percent would be excluded based on inadequate sperm count and/or inability of their samples to withstand the freeze-thaw cycle. "What's left can only be described as an elite," Kirk wrote in an email. "And they should be treated that way. Their compensation should be sufficient to require only a modest second source of income. Remember: the scheduling for twice-weekly donations leaves only one night per week for regular sex—that's something of a sacrifice (well, I can only speak for myself ☺). I think donors should be thought of more or less like monks."[51]

But of course, they could be monks with a thousand kids. And if tradition holds, Kirk admitted, most of those kids would be sired by lean monks over six feet tall. But he rejected the notion that this amounted to eugenics. "Eugenics is state-sponsored. This is what *everyone* is allowed to practice: PERSONAL eugenics. And only the mothers are allowed to choose. They get to pick the traits and the person whose genes they will help carry into the next generation."[52]

I could hear my ethicist and social-scientist colleagues groaning and gnashing their teeth. But in George's eyes, Maxey's travails were neither a deterrent nor a source of shame; on the contrary, "Kirk didn't ask to have four hundred kids. He has seen a worst-case scenario. He might not be perfectly innocent in all of this, but he has seen, firsthand, genetics gone awry. I think having people who've lived through failure is very important for the early stages of this project."[53]

"Since I'm pretty famous, I'd rather keep my name and email address private to all but the PGP researchers for now," read the email.[54]

For more than a year, none of us other than George, Jason, and the IRB knew who PGP Participant #6 was. In mid-2007, I asked George exactly what the deal was with Mr./Ms. X. "He/she doesn't want to participate in any of our discussions or with any of the press," he said.[55]

"Did you have any qualms about this person's reticence?" I asked.

"I wasn't wild about it, but . . . we should expect this to happen. Now I think it's great," he said.[56] As usual, the Churchian view was sunny. But he was not completely convincing.

I pressed him on #6's identity. Based on his/her anonymous responses to the initial questionnaire George had sent around, I was sure s/he was a woman, though I couldn't say why.

"It's a man," said George. "Just don't ask me his zip code."[57]

I wondered why this guy would want to participate in such a public project and still remain anonymous. How could he expect to have his cake and eat it, too?

"He has dealt with the press quite a bit—he's been there and done that. It's not exciting to him and he doesn't want to spend any more time on it, especially since this isn't really his party. But he's certainly interested in this topic."[58]

I thought this playing hard to get was maybe a bit pretentious, but it also lent the proceedings an air of mystery. I was intrigued. Who was this famous man who, according to his email, was a Harvard professor, hated chocolate, and didn't really want to meet the rest of us?[59]

"I've got it!" I said to George one day. "It's Alan Dershowitz!"[60] I had been thinking this for a while and was hoping it wasn't true: I found Dershowitz to be an insufferable gasbag—even in those instances when I agreed with him—and not someone I wanted to represent the PGP. It was bad enough when he was asked to speak for the Jews. Now he would be holding forth on genetics?

George laughed. "Well, I suppose it could have been him, since I know Alan fairly well. Good guess . . . but no."

And then he told me who it was. A few months after that, I met Professor Chocolate Hater himself. Steven Pinker has a lion's mane of hair (a self-described member of the "Luxuriant Flowing Hair Club For Scientists"[61]), a wide jaw, and a small, shallow cleft in his chin. He is short and thin. He wears stylish boots. He speaks in a quiet, measured, slightly nasal voice with a noticeable Anglo-Canadian accent. And he has piercing, iridescent blue eyes. In 2004 he engaged in a conversation on consciousness, morality, Gödel, and objective reality (among other subjects) in *SEED* magazine with the novelist and philosopher Rebecca Goldstein;[62] not long after that the two were married.[63] He is a prolific writer on the subjects of language, evolutionary psychology, and cognition. He is the Johnstone Professor of Psychology at Harvard. The novelist Ian McEwan said of him, "I sometimes follow up, when Steve has been in public debates, transcripts of things said on the wing. And I'm always struck by the extraordinary wit and articulacy—even though he is not at his word processor—just defending his position or indeed attacking a point."[64]

One that caught my eye was in his and Goldstein's 2007 joint interview with Salon. "[T]he reason I'm not a neurobiologist but a cognitive psychologist is that I think looking at brain tissue is often the wrong level of analysis. You have to look at a higher level of organization . . . a movie critic doesn't focus a magnifying glass on the little microscopic pits in a DVD. . . . I think there's a lot of insight that you'll gain about the human mind by looking at the whole human behaving, thinking and reporting on his own consciousness. . . . It may be that the historian, the cognitive psychologist and the biographer working together will give us more insight than someone looking at neurons and brain chemistry."[65]

He was, it seemed to me, a contradiction: a cognitive scientist interested in the granular power of biology; an evolutionary scientist and Richard Dawkins (*The Selfish Gene*) aficionado who did not seem prepared to concede the Dawkinsian primacy of the gene. He was a reductionist interested in holistic analyses.

And a guy who couldn't recall wanting to be anonymous. "Maybe it was a misunderstanding," he told me. "Or I didn't have enough information. But I don't have any clear memory of choosing anonymity."[66]

Okay, whatever. At our gathering in October 2008 he was the life of the party, asking questions of the speakers and telling jokes at the press conference about his genetic predisposition to irregular menstrual periods. And he brought his usual sharp rhetorical powers and vivid prose to bear on personal genomics in a piece for the *New York Times Magazine* in 2009:

> Assessing risks from genomic data is not like using a pregnancy-test kit with its bright blue line. It's more like writing a term paper on a topic with a huge and chaotic research literature. You are whipsawed by contradictory studies with different sample sizes, ages, sexes, ethnicities, selection criteria and levels of statistical significance. Geneticists working for 23andMe sift through the journals and make their best judgments of which associations are solid. But these judgments are necessarily subjective, and they can quickly become obsolete now that cheap genotyping techniques have opened the floodgates to new studies.[67]

For all of his caveats about personal genomics, Steve, like Jim Watson, chose not to know his APOE status. When I asked him about it, he downplayed the decision. "I know that even if I found out I wouldn't jump off a bridge. Looking at the risks, I decided I have enough existential anxieties. I know I could deal with another one, but given the choice, who needs it?"[68]

Ironically, Steve's genome did indeed turn out to be a source of existential anxiety, but APOE was not the culprit. I didn't know for a while because I didn't find out about the mutation from Steve; I learned about it from an abstract of a paper that was to be submitted by George's lab describing the sequencing of the ten of us. I didn't know if I was supposed to see it or not. But I couldn't avert my eyes. And besides, the PGP was all about transparency, right? Wouldn't this all wind up on the Web in a few days anyway? The paper called Steve's mutation "the most significant finding" of the project thus far (see chapter 12). Secrets, especially when freighted with life-threatening significance, really *were* hard to keep.

The first time I met him, Keith Batchelder posed a riddle. "Who is the biggest single commercial user of Affymetrix GeneChips?"

I had no idea. "Quest Diagnostics? LabCorp?" He shook his head and smiled, knowing the answer would surprise. "Nope. Canyon Ranch!" he said. "What's that?" I asked. "You know: Canyon Ranch! The spa!"[69] I wondered where he got that tidbit from and promptly filed it away, fairly convinced it was bullshit. But some time later, Linda Avey of 23andMe told me that this had been one of her initial ideas for personal genomics in the pre-23andMe days (see chapter 5): spas were going medical and hiring doctors. People who went to spas had disposable incomes and were interested in their health . . . and their genes.

I soon learned that Keith, five feet eleven and 180 according to his public profile,[70] trafficked in factoids and observations like this. "Do you know what the most common indication for a genetic test being ordered today in a hospital is? The patient asking for it."[71] He had tiny glasses that sat on the end of his nose, a receding hairline, and a voice that occasionally cracked when he was excited or making a point. He smiled and laughed easily, even as he told me what a cynic he was. "Life is hard and then you die." He was also a bit of a restless soul: he had been a dentist, a physician, an entrepreneur, and a research scientist. What led to this trajectory? Medicine, he told me, was just a more complex version of dentistry. Once he got his M.D., he became interested in how drugs were approved and did a stint at the FDA. From there it was biomarkers and predictive medicine. When I met him he was in entrepreneurial mode, running his own consulting business in Boston, Genomic Healthcare Strategies. This seemed to fit: he had genuine zeal for talking about genes, phenotypes, drugs, business models, commercial partnerships, and PR firms.[72]

Like several other industry players I spoke with, Keith stressed how disruptive personal genomics was and would continue to be. "When I as a patient can know a significant amount of information about my health, manage that myself, and have access to other information via the Internet, then that's a revolution."[73]

But for all of his free-market preoccupations, Keith was not espousing a laissez-faire approach. When the eight of us met for the first time, he was the one who brought up the idea of a Good Housekeeping Seal of Approval for personal genomes.[74] He was prescient: New York and California would soon crack down on personal genomics companies out of fears that patients would be harmed by receiving genetic information in an unmediated way. In the wake of these enforcement actions, talk about industry standards began to percolate. "Wouldn't you like to know whether this stuff was good or bad?" Keith said. "What agency or organization will be able to tell you this? For all of its trying, I'm not sure the federal government can react fast enough. You get toothpaste with the American Dental Association symbol on it. The ADA is not a federal agency, but an organization that has managed to create a brand and a sense of confidence. It doesn't say that if you use an approved toothpaste you're guaranteed never to get cavities. But you can assume that if you use the toothpaste it meets some set of criteria—it's not gonna kill you and the quality is at some minimal level."[75]

We talked about warfarin, the anticoagulant that millions of people with blood clots, including my nephew, most often took. Warfarin interacts with a slew of other common drugs, from aspirin to statins to Prilosec. And its therapeutic window is narrow: if the dose is too high, the patient can bleed to death (it was originally marketed as a rat poison). If the dose is too low, then the clots remain unresolved. Since the mid-1960s, we've known that different people are genetically predisposed to respond to warfarin differently—these responses can vary dramatically. People with certain versions of two genes, called CYP2C9 and VKORC1, are highly sensitive to warfarin and at risk of overdosing.[76] Keith hoped that the time to make testing of these genes standard practice in patients who needed blood thinners was nearly at hand. "We know we can predict whether you're a fast or slow metabolizer of warfarin. Warfarin is a pill that patients take themselves. Diabetes patients have glucometers and do their own insulin. They don't call up their doctors and say, 'What dose should I have today?' They may consult with a physician, they may get a prescription filled, but they manage *themselves* on a day-to-day basis. Wouldn't it be cool to have a diagnostic tool that would help you manage your warfarin dosing without having to go to a coagulation clinic?"[77]

It would be, but it appeared that it wasn't time yet: Even as this book went to press, warfarin dosing based on genotype was still undergoing randomized clinical trials. Most of the literature I read emphasized caution. But not all: surprisingly, one of the voices that advocated moving forward came from Larry Lesko, head of the Office of Clinical Pharmacology in the FDA's Center for Drug Evaluation and Research:

> The question about warfarin pharmacogenetics before us now is not "is it ready for prime time?" The more important question is, while more and more studies are being planned and/or conducted, should we accept and use our current

knowledge about genetic factors to improve the quality of
warfarin initial dosing and anticoagulation in our patients.
The benefits and risks of pharmacogenetics, in my view,
favor pharmacogenetics.[78]

Lesko's paper closed with the famous quote from Voltaire: "Le mieux
est l'ennemi du bien." The perfect is the enemy of the good.[79]

Reading this reminded me of Keith. He engaged a medical information firm to set up his own personal, portable electronic health record. He
would send around electronic articles about personalized health and ask
you what you were reading and would you send him a PDF. He said he
knew a radiologist who would scan PGP participants for free. He asked
questions about the current state of the art in genomic interpretation. He
suggested the PGP talk to Procter & Gamble or Johnson & Johnson because these companies would know how much it costs to reach consumers efficiently and systematically. They would already know, he said, how
many people were willing to pay right now and how much they'd be willing to pay in a year for a predictive genetic test. Again, he was prescient.

"People are saying that the future of the genome is twenty years from
now," he told me at George's kitchen table, eyes widening. "But look at
what Medco's doing: they have grabbed onto pharmacogenomics* in a
big way. They have gone as far as trademarking 'genetics for generics'
(can you hear branded pharma groaning?) And look at what employers
are doing . . . they are looking at ways to save money with this kind of
information. I say this is happening *now*."[80]

---

* Pharmacogenomics is the science of using an individual's genetics to determine
which drug and what dose will be most effective. Medco is a Pharmacy Benefit
Manager: it manages the prescription drug (and sometimes other) part of an employer's health benefits. PBMs try to control administrative costs of processing prescription drug insurance claims and ensuring that beneficiaries are taking the right
medications for the right conditions. One can see how genetics might help a PBM
by pointing patients to drugs that are more likely to work based on their particular
drug-metabolizing gene variants.

• • •

Participant #8 we've already met: he is Stan Lapidus, founder of Helicos BioSciences, which had waded into the next-generation sequencing market unabashed, despite being late to the party with its machine and, in a show of both up-front engineering costs and unmitigated chutzpah, began by charging more than twice as much as its competitors were for their instruments. At fifty-nine, Stan was the oldest of the PGP-10. He wore a bow tie and a faded green sweater. We sat in his nondescript office at Helicos headquarters in Cambridge's Kendall Square. When I asked Stan about whether George's plan to market a low-cost sequencer was a factor in his decision to become a PGPer, he said no. "What George is doing doesn't matter. There's no doubt in my mind that his technology won't be the one that will go the distance." He paused and reconsidered. "But what do I know? I'm just another guy."[81]

What did matter to him was genomic privacy. He believed that the PGP would be "at the vanguard of this debate." Stan told me that an individual had two rights with respect to his or her DNA. One was a right to know the content of one's genome; he described this as a "natural right" that regulators had no business in curtailing. The other was the right to genomic privacy. The second one, however, he was not prepared to grant unconditionally. "If you were an airline pilot and you were susceptible to hypertrophic cardiomyopathy, I wouldn't want you to be an airline pilot," he said. "Pilots are given stress tests every six months because a lot is riding on their health. Is it fair and appropriate to step into the realm of genetic testing for people whose genetic differences can affect the lives of others? It's a complex question."[82]

Stan had googled me and read an article I cowrote with Bob Cook-Deegan about gene patenting. The subject, it turned out, was another of his fixations: genes were *discoveries*, he said, not inventions, and therefore should not be protected by intellectual property rights. Worse, gene patents would stand between the citizenry and their natural right to know their genome sequence. "Patenting discoveries makes no sense at all. I never understood it. So 23andMe can sequence this and that but it can't sequence your breast cancer genes because they're patented?* It's crazy!"[83]

Stan was on the back nine of a long career as a biotech entrepreneur: he founded two companies prior to Helicos and nurtured thirty-one-year-old wunderkind and Knome CEO Jorge Conde,[84] among others. He told me about being at Columbia University during the Vietnam War and having a physics professor who had played a cameo role in the Manhattan Project. In the mid-1940s, Stan explained, the thinking of Robert Oppenheimer and the rest of "the atom-bomb boys" never went beyond the notion that we were at war and the fear that Germany was developing a bomb; that we had to develop one first seemed obvious. But thirty years later, Stan's professor had palpable regrets about how he had spent the war. Stan had a deep-seated desire to avoid the same fate. He described a scenario in which a Slobodan Milosevic could use genetics to carry out ethnic cleansing with "pinpoint precision." He imagined Iraqis sorting Shiites from Sunnis, and the stern rabbis of his childhood using genetics as the final arbiter in deciding who was a Jew.[85]

None of this is to say that Stan was against wielding the power of genetics and genomics to bring about social change. Like Kirk Maxey, under certain circumstances he was all for it. He offered another vignette. "Suppose I were single and wanted to meet women who were college-educated, interested in the sports I'm interested in, interested in philanthropy and the humanities, scientifically accomplished, and who didn't carry the same recessive diseases as me. Well, the last one can be done. Doing the matching can occur semi-anonymously and in such a way that you simply won't meet the woman that you 'shouldn't' meet so she won't break your heart. This would be useful for things like Facebook and Yahoo! Dating."[86] (Indeed, another PGP-10er, Steven Pinker, had helped

---

* This did not stop 23andMe from returning results on a subset of patented breast cancer mutations, however; the company began offering these results in February 2009; http://www.genomeweb.com/dxpgx/23andme-adds-brca-breastovarian-cancer-testing-service.

found a company aimed at young couples based on broad carrier screening for dozens of genetic diseases.)[87]

But doesn't that add up to eugenics? "There is a eugenic by-product," Stan conceded, "but the immediate impact on families would be to help assure that they won't have a terrible outcome."[88]

I was reluctant to get into this debate for the umpteenth time. And anyway I had my own problems (I know: spoken like a true personal genomics enthusiast). Just before my conversation with Stan, my phone rang. It was Ann; her voice quavered and I could tell she'd been crying. Her CA-125 was high. *Again*. CA-125 is a protein that's detected in blood, elevated levels of which can be associated with ovarian cancer. Emphasis on "can." CA-125, a gynecological oncologist told me, is a shitty test, subject to both false positives and false negatives. In other words, a woman may have no detectable elevation in her CA-125 and still have deadly ovarian cancer. Or she may have a CA-125 that's off the charts and therefore be convinced she will die imminently, only to discover that she's fine and her CA-125 was elevated for some other completely benign reason such as endometriosis or even normal menstruation.[89]

So why use CA-125? In short, because it was the best we had. Unlike skin or blood, ovarian tissue is difficult to access. Women with ovarian cancer often complain of nothing more than vague abdominal symptoms. Ovarian biopsies aren't done because they pose a risk of spreading a cancer. The only way to diagnose the disease definitively in a woman is to open her up, something that obviously cannot be undertaken lightly. The net result is that while ovarian cancer is relatively rare, it's the deadliest tumor a woman can have. The development of a reliable, noninvasive diagnostic is among the holy grails of gynecological medicine, to say nothing of a potential gold mine to its inventor.

With some trepidation, I unburdened myself to Stan, a guy I'd known for all of half an hour but most of whose professional life had been spent developing cancer diagnostics. He responded by switching into problem-solving mode. Stan swiveled around in his chair and began scrolling through his email, looking for messages he'd received from a scientist at Yale who was working on a novel blood-based ovarian cancer diagnostic. The test, called OvaSure, was purported to be able to catch 95 percent of cases.[90] Ninety-five percent! The rights had been sold to LabCorp and it was due to go on the market soon. In my unhinged mind I had already decided that Ann had ovarian cancer and that she would be among the first women to avail themselves of this new miraculous tool. She too would be an early adopter!

Wrong and wrong.

Ann's CA-125 levels returned to normal (whatever that means) in subsequent months; it danced around but was never high enough to alarm her gyn-onc. She had a couple of fibroid cysts but nothing re-motely malignant. The gyn-onc we saw, a short and friendly man with wire-rimmed glasses, shrugged and repeated what he'd said at Ann's last appointment: "CA-125's a terrible test and it causes a lot of needless anxi-ety." I asked him about OvaSure; he made a face and shook his head. "Those guys still have a *long* way to go."

He was probably right about that. LabCorp brought OvaSure on the market as a "home-brew" test offered by a single lab. Under current regu-lations, this meant that it did not need FDA approval. The FDA, however, saw things differently. It said that 1) the test was developed at Yale and not at LabCorp and therefore was not home-brew and could not circumvent FDA regulation; and 2) the agency needed to be convinced that the test worked. In the interest of remaining in the FDA's good graces, LabCorp quickly yanked OvaSure from the market.[91]

A few months later I learned that a friend and former teacher, Sheila Schwartz, died of complications of ovarian cancer. Like most women with the disease, she was diagnosed when the cancer was already at an advanced stage, though she fought it for many years. She learned later that she carried a BRCA2 mutation. She was fifty-six.[92] Not long after that, my cousin was diagnosed with ovarian cancer. She was forty-two.

"We stop where the science stops."

This is what Participant #9 told me about the nutrigenomics company she cofounded, Sciona. For Rosalynn Gill, this statement had become something of a mantra.[93] In 2006, Sciona was one of four companies hauled before Congress following an investigation by the U.S. Government Accountability Office. The GAO suggested these companies misled consumers by selling tests that offered predictions based on diet and lifestyle surveys coupled with genotype information. The chairman of the committee presiding over the congressional hearings, Senator Gordon Smith, Republican of Oregon, derided the companies' products as "snake oil." In full indignation mode, he lectured them: "I don't want consumers preyed upon in such a manner."[94]

Rosalynn called it the most terrifying day of her life. And then she rolled her eyes.[95] The GAO, she explained, set up a sting of sorts, in which it submitted fourteen lifestyle questionnaires from fourteen fictional people using only two DNA samples, one from a nine-month-old infant and the other from a forty-eight-year-old man. The GAO's indictment of the companies' practices was based in part on the fact that each of the fourteen phony customers received an individualized set of recommendations despite representing only two distinct DNA samples.[96] Rosalynn thought this was actually a good thing. "I was relieved!" she said of the different recommendations. "This meant that my software was working as it should and taking into account customers' lifestyle data."[97]

The Sciona protocol was similar to other direct-to-consumer tests: a customer spit in a tube, sent it off, and in a few weeks got results. The company farmed out its genotyping to Connecticut-based Clinical Data, which returned genotypes on variants in nineteen genes associated with nutrition and health. A customer, for example, might be found to have a particular DNA variant in a gene linked to poor absorption of calcium or vitamin D and therefore possibly at risk for developing osteoporosis. Based on what the person reported about his/her diet, Sciona may have advised an increase in vitamin D, calcium, omega-3 fatty acids, and/or exercise in order to promote "bone health." Or if someone had a certain variant in her MTHFR gene, she might be advised to eat a diet richer in folic acid and/or take B vitamin supplements in order to reduce her homocysteine levels (elevated homocysteine is a significant risk factor for heart disease).[98] Senator Smith referred to this type of activity as "diagnosing disease."[99] Rosalynn objected—there was a difference between risk factors and actual disease. "I said, 'Wait a minute! I'm not promising to reduce the risk of disease. I'm not diagnosing disease. We're talking about how these genes can affect the levels of things like homocysteine and cholesterol.' "[100]

When we spoke at the second PGP gathering, she referred to herself as a "healthy forty-seven-year-old woman" and therefore was not expecting much from her genomic consult. A molecular biologist by training, she was tall, blond, slender, and smiled a lot. She was divorced and lived with her children in Colorado, where Sciona was headquartered. The day before the PGP participants met, she had run a half-marathon. She saw personal genomics as a tool for self-improvement. "Current models are not working for nutrition education. If we can help someone to make the right choices, to eat more healthfully, then I think that that demonstrates the greatest utility of what we're doing. I'm really interested in making a difference in people's everyday lives."[101]

She had done both Navigenics and 23andMe. But, she said, because Sciona was the first company to give genetic data to consumers, she was more interested in the way personal genomics companies *presented* their results than in the actual results themselves. She thought that by ignoring environment, diet, lifestyle, and family history, the risk estimates she got back from Navi and 23andMe were close to meaningless. She sounded like many of the scientists and clinicians I'd spoken to, one of whose main messages was essentially "genes, shmenes—where is the phenotype?" She ignored the 23andMe and Navigenics disease risk numbers. On the other hand, learning her ancestry via 23andMe was fun, she said: her Central European maternal lineage, for example, was a surprise. "I tell my friends I'm descended from Attila the Hun."[102]

Reading over the congressional testimony on the GAO report[103] and an investigation conducted by the United Kingdom's Human Genetics Commission,[104] I found it hard to muster much outrage toward Sciona. Maybe that was due in part to Rosalynn's skillful politicking and infectious charm ("If my weight goes on the Internet, George, I'm out of here!"[105]) or the even-keeled way she stood up to a bullying senator. The governmental wrath the company had incurred in two countries seemed a little disproportionate to its so-called offenses.

At the same time, I couldn't see myself spending $250 on Sciona's Mycellf Program[106] in the hopes of "personalizing my diet." Just as Rosalynn criticized the personal genomics companies for not collecting trait information, there was no reason to believe that the genetic bases of traits Sciona reported on, such as weight and cardiovascular health, were not a hell of a lot more complicated than what one could glean from typing twenty-odd genes, many of which were poorly understood, with or without lifestyle surveys. If, when talking about nutrigenomics, we were stopping where the science stops, then we were probably not going very far. Not yet, anyway.

But Sciona wasn't going to see the future, no matter when it arrived. When the economy tanked in 2008, it could not raise enough capital to stay afloat. By the following summer it had closed its doors; its assets were sold soon thereafter.[107] Rosalynn found a position with Ipsogen, a French start-up focused on the molecular diagnostics of cancer.[108]

James Sherley was the last of the ten and the only African-American. I mention his race because it is relevant to our story. George wanted to include one or more minorities in the initial round of the PGP for the sake of both genomic and cultural diversity.* I had recently interviewed Rick Kittles, a prominent black geneticist, and suggested to George that he might be a candidate. George corresponded with him, and he showed

---

* While our genomes are 99.9 percent identical to one another on average, the 0.1 percent still adds up to 3 million differences. And because human populations were geographically separate for long periods of time, it is possible to make fairly precise estimates of someone's continent of origin based on a few thousand DNA markers. Indeed, this is how 23andMe (see chapter 5) and companies like Family Tree DNA assess their customers' origins.

some interest, but ultimately decided not to do it (see chapter 13). In fact, many African-Americans—even research geneticists—were wary of biomedical research. It's easy to understand why. The Tuskegee "studies" of untreated syphilis in African American males dragged on for decades.[109] In the 1970s and '80s, the U.S. Air Force Academy rejected 143 black applicants who were merely carriers of the sickle cell trait.[110] And in the consciousness of many people of color, DNA connotes not wonder, but the sinister side of forensics: dragnets, lack of due process, and the unchecked power of government and law enforcement.[111]

James Sherley's view of DNA and the human genome, however, was that of a scientist. When the PGP-10 gathered for the second time, he marveled at the idea that, with our cell lines in a public repository, we had each been "immortalized,"[112] just like Henrietta Lacks, the dirt-poor African American woman for whom the HeLa cell line had been named.[113]

He could talk for hours about stem cell biology and how our genomes programmed cells to carry out various functions. That said, he was hardly disengaged from matters of race. When George recruited him, Sherley was busy fighting his own public battle against what he perceived to be racial discrimination. Sherley came to MIT in 1998 as a junior faculty member in the new Department of Biological Engineering. In retrospect, he told me, things went wrong at the beginning. He was the first investigator hired into the new department, something he said the BE division head, Douglas Lauffenburger, never acknowledged, instead claiming that he was hired into the Division of Toxicology. Some thought Sherley was just being petty, but for him the symbolism mattered, especially since he would remain the only African American in BE for at least nine years.[114]

Sherley said he had less laboratory space—the coin of the realm in university science—than his other junior colleagues. And he learned that he had been hired into a targeted minority slot, something he said that no one at MIT had ever felt the need to mention to him. This mattered because MIT did not allocate dedicated space resources to targeted minority hires; they had to make use of space that was already available. According to Sherley, then–MIT provost Robert Brown declared that he would not give space to either women or African Americans.*[115]

In 2004, Lauffenburger informed Sherley that he would not advance his name for tenure. At the time, Sherley said, Lauffenburger also made a cryptic remark: he *knew* that Sherley's race was going to be a factor in his tenure decision. Sherley didn't know what that meant, but the fact that it was said at all was enough for him to demand an investigation. Sherley also believed there were conflicts of interest at work in his case.[116] Lauffenburger was married to Linda Griffith, another MIT stem cell biologist who had had sharp disagreements with Sherley on matters of science, although she had been principal investigator on grants that had partly funded Sherley's research. Nevertheless, a letter written on Sherley's behalf described their relationship as "openly contentious."[117] For this reason Sherley thought Lauffenburger should recuse himself from all matters relating to the tenure decision; Lauffenburger did not. In fact, he solicited a letter from Griffith to be included in Sherley's tenure dossier.[118]

In 2001, a month before 9/11, President Bush banned all federal funding for embryonic stem cell research.[119] In the 2004 presidential election, stem cell funding became a major issue. Michael J. Fox and Christopher Reeve agitated for public support.[120] Molecular biologists, too, stormed their state legislatures, which began allocating billions for homegrown embryonic stem cell (ESC) research programs.[121] Sherley was not among those lobbying for ESC funding, however. On the contrary: the son of a Baptist minister, James believed that embryonic stem cell research resulted in the destruction of human life. "Like thousands of other multicellular organisms on this planet," he told *Celebrate Life* in 2007, "human beings start life as a single-cell embryo, the product of the union of a complete human genome and the programming cytoplasm of a human egg. This union occurs at fertilization. . . . I challenge the promoters of

---

* MIT faced a minor insurrection by female faculty members in the 1990s, which resulted in an acknowledgment of gender discrimination and a concerted effort to address it.

human embryonic stem cell research to justify why another human embryonic life is less worthy than their own. . . ." He would work only on adult cells.[122]

Given the tenor of the debate and what was at stake in the early 2000s, it seems unlikely that his position on stem cells could have won Sherley many friends in Biological Engineering at MIT. He wouldn't venture a guess as to how important his stance was in the decision to deny him tenure and in the ensuing kerfuffle; he would say only that he would be naïve to think that it didn't play a role.[123]

In September 2006, more than a year after his tenure decision, Sherley won the prestigious Pioneer Award from NIH for his work on adult stem cells. The $2.5 million award supports "exceptionally creative scientists who take highly innovative approaches." That year just thirteen were granted.[124]

But this development neither bolstered his case for tenure nor otherwise improved his position in the eyes of the university. In subsequent months relations between MIT and Sherley deteriorated. In December 2006, he wrote a letter to the campus newspaper in which he said, "I will either see the provost* resign and my hard-earned tenure granted at MIT, or I will die defiantly right outside his office."[125] He refused to communicate with the provost, who he said mishandled the grievance process.[126]

On February 5, after two bowls of Chex cereal, James Sherley began a hunger strike and a vigil outside the offices of MIT's top administrators. His doctor, who helped him prepare and visited him throughout, said, "I hope you survive this." He responded: "Well, if I don't you'll be the first to know."[127]

The following day, a letter from a group of MIT professors, led by leftist linguist Noam Chomsky, wrote to the campus paper on Sherley's behalf questioning the integrity of the grievance process. George Church was among the signatories.[128]

---

* The provost by this time was L. Rafael Reif.

After twelve days and twenty pounds, Sherley called it off. The provost said there could be no mediation while a hunger strike was going on. With the help of some emissaries, Sherley and MIT released concurrent statements. MIT "deeply regret[ted]" that Sherley's experiences led to his fast. His protest "focused attention on the effects that race may play" in the career trajectories of minority faculty. MIT said it would "continue to work toward resolution of our differences with Professor Sherley." For his part, Sherley said he was breaking his fast "in celebration of the attention that has been brought to bear on issues of equity, diversity and justice at MIT" and elsewhere. He also said that his demands were "still on the table."[129]

It soon became obvious that those demands would not be met. At the end of March his colleagues in Biological Engineering released a statement saying that the tenure process was a "fair and honest" one; it suggested Sherley's publication record was not up to snuff and noted that MIT could not consider work Sherley had performed prior to coming to MIT.[130] After an exploratory process looking for tenure opportunities across campus, the provost made it clear that tenure was off the table in any department at MIT and that Sherley's appointment would end on June 30.[131] A colleague representing Sherley characterized the provost's proposal for mediation as a "notice of eviction."[132] By then Sherley had long since stopped communicating with the provost. In May the provost released a statement saying that MIT had never agreed to review Sherley's tenure case again, or to conduct any review of his allegations about the grievance process.[133]

June 30 was a Saturday. When Sherley tried to gain access to his lab on the following day, he found that his keys no longer worked. He emailed the president of MIT, worried for the viability of his cell lines and mice.[134] The provost told the campus newspaper that Sherley had declined earlier opportunities to avail himself of university assistance. Consequently, MIT had decommissioned Sherley's lab in accord with "all applicable rules and regulations."[135]

My first encounter with James was at the pseudo-vegan PGP barbe-cue at George's house just weeks after James had been locked out of his lab.[136] We stood at the bottom of the stairs in George's artsy and plantsy Brookline bungalow and talked quietly over beer; James still seemed shaken by all that had happened. He paused frequently to push his glasses up his nose. He was not a skinny man; during his hunger strike the Harvard newspaper, among others, had ridiculed his weight (and his cause) and suggested that a hunger strike would be good for his health.[137]

But if I were expecting an angry, militant black dude, I was to be disappointed. James spoke with passion about his cause, regret about what the ordeal had done to his family and friends, and anguish about the time it had taken from his science. He also laughed easily and told me he was relishing the time at home with his kids while he looked for a new job.[138]

When we met in a Brookline coffee shop the following spring, he was less preoccupied, though he assured me he was not at peace. The Equal Employment Opportunity Commission, with whom he had filed a discrimination suit, had not found in his favor;[139] this discouraged him. And he was still angry with some of his allies, who he felt did not call out the MIT provost with sufficient vigor. He blamed himself for the same lack of will: by not resuming his hunger strike, he said, he felt he had let MIT off the hook (although he admitted that his decision also spared his wife and eleven-year-old daughter further anguish; his daughter was especially concerned that her father really would "die defiantly").[140]

By the second PGP gathering his professional life had begun to settle. He was now a senior scientist at Boston Biomedical Research Institute in Watertown. The director was already familiar with James's science. But did BBRI flinch at his stance on ESCs or his MIT bag-gage? "There was one question during my interview: 'How do you feel about ESC research at BBRI?' My response was truthful: I'm not going to get in the way of any programs that involve ESC. But I'm also not going to participate in them. I will continue to be an outspoken advocate on this topic, but I've never interfered with anybody's research."[141]

Did James Sherley deserve tenure? I have no idea. Most junior faculty at MIT *don't* get tenure; this, after all, is why it's MIT, a truly elite institution. Had he won the Pioneer Award two years earlier, before the decision and the subsequent escalation, perhaps things might have gone differently for James. And it's not hard to see how he might have hurt his own cause with his demonstrativeness, his high-profile martyrdom, and his pointed and often public accusations. But my guess is we will never know because I suspect the process was doomed from the beginning by miscommunication and a lack of transparency on the university's part.

In January 2010, MIT published a comprehensive study of racial diversity among its faculty[142] (disclosure: my brother Josh is an economist at MIT and was on the committee that crunched the numbers for this report). While the Sherley incident was mentioned just once in the 156-page report, I found it hard to believe it was not a major impetus for MIT's decision to reflect upon its own minority hiring and promotion track record. Among the findings: 74 percent of Caucasian assistant professors were promoted to associate professor without tenure (a crucial step on the path to tenure), while only 55 percent of black and Hispanic assistant professors were promoted to associate—a statistically significant difference. Of course, only 6 percent of MIT's faculty was black or Hispanic to begin with (a fraction comparable to Harvard and Stanford).

Perhaps future African American scientists—and those who make the tenure decisions affecting them—will benefit both from the report and from the Sherley episode itself. That might be the best we could hope for.

After the tenure drama and the hunger strike and the protests, George hesitated to ask James to participate in the PGP. Not because he shied away from controversy (he would not be George Church if he did that), but because he couldn't imagine his friend and collaborator would want to undertake yet another high-risk, high-profile controversial venture. He was wrong.

"I kept wondering, 'Why isn't George asking me about this?'" said James. "When he finally did I was like, 'What took you so long?' I'm interested in being fair in disclosure and thinking about how to build research programs that can actually be responsive to the needs of participants in the study."[143]

Eventually James and I got around to discussing our genomes, and the aggrieved academic instantly morphed into the awestruck nerd. James was not terribly interested in genes per se (cell biologists often operate at a higher, more systemic altitude than molecular biologists).[144] At one point during the initial PGP meeting he said he doubted he would even look at his own genome, though he later backed off from that statement. But he said he was much more curious about his "origins of replication," the sequences where DNA initiates the process by which it copies itself; there is some controversy about where they can be found in the genome.[145]

James's real passion was stem cells. Stem cells are undifferentiated cells. We all begin existence as embryonic stem cells. They are valuable because their fate has not yet been determined; they have not committed to becoming neurons or T cells or muscle cells or melanocytes. Their value in medicine, so far largely unrealized outside of leukemias, is that they have the potential to become almost any type of cell (for this reason they are referred to as pluripotent). They could become brain cells in a Parkinson's patient or pancreatic islet cells in a diabetic. If we want to get the greatest number of stem cells with the broadest range of possible fates, embryonic stem cells are ideal. But many, like James, are not comfortable with destroying them because they see them as human life. These scientists have therefore focused on reprogramming *adult* cells: taking a skin cell, for example, and "dedifferentiating" it back to a stem cell state where anything, or many things, are still possible. This is called induced pluripotency: taking a differentiated cell and turning the clock back so that it behaves like a stem cell.[146] Sherley's lab had helped figure out how to do this and how to get these cells to divide in culture so that they generated a renewable supply of themselves.[147] The problem was that so far it was a terribly inefficient process. Making it more efficient was one of his and George's goals.

James was excited about stem cells for other reasons. One was the biology they might teach us. Another was their potential predictive value. "Cancers may be stem cell diseases. Asthma, too. Multiple sclerosis may be a stem cell disease. Changes in the genome of undifferentiated cells may predict [health outcomes] because those are the cells that are constantly renewing."[148] By the time we reach the fetal stage (the ninth week of pregnancy), our developmental potential has been written into most of our cells. "If you could only make one gene expression pattern in a person's life and see which genes are turned on and off, the fetal period would be the time to do it."[149]

Different genes are turned on and off at different levels at different times in different types of differentiated cells; this is reflected in the different levels of RNA one finds for the same gene in different cell types at different stages of human development. This is what makes a neuron a neuron and not a hair follicle. James and George wanted to conduct an "organ recital"—that is, they wanted to measure gene expression in each of the 200+ different types of cells in the human body.[150] For blood, hair, and umbilical cord, that was pretty easy. Buccal cells could be scraped from the inside of the cheek with no problems. Skin biopsies hurt but were tolerable. But once you got into muscle and liver biopsies, to say nothing of brain tissue, then access became a real issue. So why even bother? Because if we really wanted to understand human beings at the molecular level, organ recitals were a necessity. "We're more than our genes," explained James. "We're the *expression* of our genes. Looking at differences in expression is going to be much more informative than genotype."[151]

It was 1:30 P.M.; we had been sitting in the coffee shop for more than two hours. Outside we could still see our breath; we had to step over the snowbanks to cross the street but the sun was shining brightly in Brookline. As we walked to James's car, I asked whether he thought the PGP could really make a difference. Wasn't personal genomics a luxury? "Listen, there are lots of other needs in society. Should we be spending this money on building houses for people? I don't know. I certainly think and worry about it. But as a scientist I feel that knowledge shared is the best thing we can have. What *should* we be gathering knowledge about? Understanding how we work, how we function, how we grow up and smile— that will be difficult to do. But it's hard to argue that that isn't important somehow."[152]

7

"It's Tough to Guard Against the Future"

Thanks to the late, great, and supercheap Skybus Airlines, Ann and I could afford to bring the kids to what had become, after only one visit, my favorite meeting: Advances in Genome Biology and Technology, in Marco Island, Florida. As I intimated in chapter 5, AGBT is to DNA sequencing what Macworld is to all things Apple: a highly anticipated annual unveiling of sleek new toys, replete with the requisite backchannel discussions, competitiveness, fire hoses of data, and more than a little showmanship, all at the plush Marriott resort right on the beach.

AGBT could be counted on for theatrics (Pacific Biosciences, still more than two years away from launch, sponsored fireworks over the Gulf) and rumor-mongering ("So and so's new machine is having problems"). But by 2008, the next-generation sequencing field had gotten more crowded,[1] with Illumina having overtaken first-to-market 454 and ABI trying to play catch up after launching only a few months prior to the February meeting.[2] Half a dozen other companies claimed to be "close" to bringing instruments online. Attendees heard big-picture, crystal-ball talks about "the future of DNA sequencing" and more narrowly focused presentations with sexy titles like "Respiratory Bacterial Pathogens Utilize Polyclonal Infections and a Distributed Genome as Population-Based Chronic Virulence Traits." I confess I skipped that one for some quality time in the Tiki Pool and a few drinks with umbrellas in them.

George was not there in the flesh, but certainly was in spirit. On a late afternoon inside a rented suite at the Marriott stood his lab's crowning technological creation (at least for the moment): the Polonator. I had already seen the hastily assembled marketing piece, which was adorned with a picture of a large blue box that had been slightly warped in kind of a Daliesque way, a nice design touch and appropriately Churchian. The splashy-cum-nerdy brochure announced the machine's arrival:

> Dover, in collaboration with the Church laboratory of Harvard Medical School, introduces the Polonator G.007, a revolutionary approach to second-generation sequencing. The Polonator G.007 is a completely open platform, combining a high performance instrument at a very low price point . . . [Users] are totally free to innovate; all aspects of the system are open and programmable. . . .
>
> [The Church lab's] vision, as expressed in the Personal Genome Project, the development of the Polonator, and their recent formation of the PGx team [that will compete for the X Prize], is quite simple: to deliver the benefits of second-generation sequencing to the largest possible base of potential users, as quickly and efficiently as possible.
>
> . . . For those intrepid souls willing to slip into the driver's seat, the Polonator is completely open and at your disposal. Beyond buckling up, our only request is that you respect the open nature of the Polonator system, and promptly publish (or better yet communicate immediately via our user community forums) any enhancements that

you might develop. It is through your creativity that the
Polonator system will evolve.[3]

The Polonator would be more than just a sequencer; it would be a
philosophy—a way of life. Like the PGP, this thing was not for the timid.
Why this approach, George? "For anything you do on government grants,"
he said, "open-source is a good idea. Also, I survived the Applied Biosystems monopoly; I always felt it was too hidden and it was too stifling. I
always felt like, 'Gee, if somebody had the resources, wouldn't it be great
to just have all this stuff in the open?' I think that feeling comes from having benefited from open-source software myself. And it also seems to be
consistent with what we're doing with respect to the ELSI [ethical, legal,
and social issues] aspects of the Personal Genome Project. We are trying to
be transparent in every way."[4]

Applied Biosystems was the company that won the race to develop
automated sequencing in the 1990s. By convincing bad-boy scientist
Craig Venter to undertake a private effort to sequence the human genome and compete with the government's Human Genome Project, ABI
became the arms dealer that catered to both sides of the war. Venter and
the NIH-sponsored public effort each bought hundreds of ABI's instruments at three hundred thousand dollars a pop. And with the company's
model 3700 (and its successor, the 3730) installed as the industry standard,
ABI could then charge lots of money for reagents: proprietary molecular
bullets for the company's high-tech guns. It was a brilliant move and it
helped to keep ABI's balance sheet deep in the black and its machines
entrenched in hundreds of molecular genetics labs for the better part of a
decade.[5] But now obsolescence was in sight: the wealthier labs had already
begun to switch to the new platforms and draw down sequencing on
the old machines.[6] In 2009, I saw a used 3700 on eBay for about $3,500,
shipped. In another auction, its predecessor, the 377, was sold for a winning bid of $99.

In the suite at the Marriott I helped myself to a glass of red wine
and inspected the Polonator up close. Even someone who had no idea

what it was had to be impressed with the aesthetics. It was electric blue, a much bolder color than any of the competition's wares. Its glass door was flanked above and below by bands of orange racing stripes. Inside were all of the usual moving parts characteristic of a next-gen sequencer: a robotic platform to cradle the flow cell—that is, the small piece of glass that held the DNA to be sequenced; a charge-coupled device camera to record the images of each base after it was incorporated; and lots of tiny capillary tubes that moved enzymes and other reagents in and out of the flow cell. Next to it was a computer whose job it was to instruct the Polonator. And lo and behold, this Polonator was actually on! Its lights were flashing and its parts humming as they moved stuff from one place to the other. If this thing worked, I thought, I might actually get my sequence someday.

The man who had taken the technology from the Church lab and turned it into an exemplar of sleek design was Kevin McCarthy. Were one to draft a prototypical George colleague from the technology sector, Kevin would probably be pretty close to the final iteration: a tousled mop of gray hair, skinny with a buttoned-down shirt, oversize wire-rimmed glasses. An engineer not entirely comfortable as pitchman but full of ardent belief in his product. An inventor.

A year earlier, McCarthy, who'd been put on to the Church lab by a sales rep, was being escorted to Harvard Medical School to meet with George's team. But his coworker/driver got lost in the medical school maze. By the time they made it to George's office, they had all of fifteen minutes to make their pitch. "And it wasn't like they were dying to see us," McCarthy said.[7]

At the time the Polonator was much closer to a science fair project than a commercial product. When I first visited, "Polonator Central" was still a small windowless room within the sprawling Church lab. Inside a boom box throbbed with the Rolling Stones' *Steel Wheels*. The first thing one noticed was the temperature: 18° C, or about 64° F. Ligase, the enzyme used to stitch DNA together in the "Polonation" process, liked it cool; early versions of the Polonator had no onboard refrigeration (or onboard much of anything), so Church protégés Jay Shendure and Greg

Porreca had to keep the ambient temperature down. The room was littered with tangles of black wires connecting microscopes to computers on
stainless steel shelves. Space was so tight that the keyboards dangled verti-
cally over the sides; Rube Goldberg would have been proud. Each one of
these makeshift arrangements had been named after a character from *The
Simpsons*. Marge, I noticed, was in the midst of running a batch of samples.
On the screens were thousands of white dots, like stars on an especially
clear night. These were polonies—short stretches of DNA that had been
amplified millions of times.

"I wandered into the lab and saw some interesting things," McCarthy
recalled diplomatically. "I said, 'I can do a lot of these things better than
this stuff you've cobbled together.' It began with a few core components
and just kept getting bigger. I said, 'Would you like some metal to connect
all this stuff?' Finally it got to the level of, 'How about we just make the
whole thing?' They were like, 'Yeah! Let's go!' "[8]

Back in the main ballroom, Baylor's Amy McGuire was making a presentation on the ethical challenges of personal genomics. We were living
in an exciting time, she said, but the excitement was tempered by worry.
All of the events of the last year or two—the publication of individual
genomes, the X Prize, NIH grants meant to hasten the arrival of the
thousand-dollar genome, genome-wide association studies, and yes, the
PGP—raised some ethical issues that for the most part had not been in
play before the digital and genomic ages. Among the biggies:

- Did researchers believe their own informed-consent forms
  and did subjects understand them?
- Should research results be returned to research subjects?
- How should investigators share data, and with whom?

Projects like the PGP raised the larger question: Can participants
consent to uncertainty? To paraphrase Donald Rumsfeld, could we accept

not knowing what we did not know? Clearly the ten of us thought we could; we also thought that anyone who was comfortable with the unknown should be allowed to assume those risks. NIH, on the other hand, was clearly *not* down with the idea; governments and regulatory agencies liked certainty. McGuire did not take a stand—"All of these approaches have merit," she said.[9] But unlike the rest of us, she had actually walked the walk: she was the one who had navigated the ethical path for the release of Jim Watson's genome to the world. Watson was, for the most part, like the PGP-10: he was prepared to let it all hang out—to deposit his genome into a public database and let the world have unfettered access to his gene sequences and the rest of his DNA.

With one exception.

In the 1990s, scientists at Duke University (disclosure: my employer since 2003) discovered that the apolipoprotein E (APOE) gene was a major risk factor for garden-variety, late-onset Alzheimer's disease that some 5 million Americans are living with (there are rare, purely genetic forms as well). One copy of the APOE4 version of the gene put you at threefold higher risk of developing Alzheimer's. Two copies and you were really in trouble: by age eighty-five, more than 50 percent of people with two copies developed Alzheimer's.[10] Would people really want to know if they were at such high risk for a disease they could do almost nothing to treat or prevent?

Watson had made his stance clear in the press and also during our interview at Cold Spring Harbor. "My Irish grandmother died of Alzheimer's at eighty-three," he said. "I don't want to worry that every lapse in memory is the start of something. I'm not afraid of the future, but I don't want to know. Of course, I could be homozygous APOE4 and still not get Alzheimer's, so . . . it's complicated."[11]

Indeed it was. After McGuire's talk, questioners from the audience lined up in the aisles. Among them was a fresh-faced young man in shorts and a T-shirt. This was Mike Cariaso, who, according to a speaker bio I read for a later meeting, "enjoys travel, skateboarding, reading genomes, and programming in Python."[12] When he got to the mic, Cariaso informed

McGuire—and the rest of the audience—that it was possible, and in fact quite easy, to infer Watson's genotype using his available DNA sequence data from one and/or both sides of the APOE gene.

Geneticists have given this phenomenon the rather unwieldy name of linkage disequilibrium. What it basically means is that even though parts of each parent's set of chromosomes* get exchanged when sperm and egg come together (recombination), the process is not entirely random: two genes that are right next door to each other on a chromosome tend to travel together through the generations. Genes that are on the same chromosome but farther apart are more likely to be separated during recombination—they are less tightly linked. In order to prevent Watson's APOE status from being known, Baylor had redacted or "scrubbed" his APOE sequence and some amount of DNA on either side. But it wasn't enough. Cariaso was still able to use the more distant sequence: by knowing which versions of the SNPs on either side of APOE Watson inherited, Cariaso could check the genome databases and see what version of APOE tended to travel with the more distal parts of the chromosome that hadn't been scrubbed. This type of deduction—comparing what's known in other cases to an unknown case like Watson's—is how linkage disequilibrium works.

"Watson may not know his APOE genotype, but I do," Cariaso told McGuire in front of the stunned crowd. "And if anyone else wants to know, the information is still on the [National Center for Biotechnology Information] server."[13]

He returned to his seat. Someone in the audience, an aghast and agitated Baylor genomicist I imagined, her face pale, marched up to him and began firing questions, each of which Cariaso answered with quiet confidence. The facts were inescapable: The Nobel Prize–winning and DNA-discovering source of the second completely sequenced human genome had asked that of his 20,000+ genes, sequence information from just one

---

* Recall that chromosomes are the twenty-three pairs of vessels that carry our DNA as tiny packages and are found in the nucleus of every cell.

lousy gene—*one!*—not be made public. This task was left to the molecular brain trust at Baylor University, one of the top genome centers in the world. Its mission was to keep secret a single genotype from a single gene. But the Baylor team was outfoxed by a thirty-year-old autodidact with a bachelor's degree who preferred to spend most of his time on the Thai-Burmese border distributing laptops and teaching kids how to program and perform Google searches.

If there were any remaining doubts as to the relatively easy availability of Watson's APOE status, they were erased a few months later when Australian researchers came to the same conclusion as Cariaso. The title of their paper in the *European Journal of Human Genetics* said it all: "On Jim Watson's APOE Status: Genetic Information Is Hard to Hide."[14]

After I got home I asked Mike via email what he made of the minor shitstorm he had started at Marco Island. He wrote back: "The ethical conundrum is: What did Watson intend *not* to know? Was it: 1. 'Don't tell me my APOE sequence'; 2. 'Don't tell me my ApoE4 [trait] status'; 3. 'Don't tell me anything that might reveal my ApoE4 [trait] status'; or 4. 'Don't tell me anything that predicts Alzheimer's?'

"Number 1 and Number 2 were addressed. Number 4 is impossible, since it's based on what we might discover tomorrow. Given the best data we have today, we know that Number 3 wasn't covered due to the high linkage disequilibrium with a distant neighbor [on the same chromosome]. If they'd scrubbed APOE [plus another] 30,000 base pairs on either side, then they would have covered what we know today. But that doesn't mean tomorrow we won't learn a new way of determining [APOE genotypes] from some sequence 50,000 base pairs away or even on a different chromosome. It's tough to guard against the future."[15]

Mike and his friend Greg Lennon run SNPedia, a wiki-based website that is in some ways the do-it-yourself version of 23andMe. Sort of. Mike, Greg, and anyone else who wants to can "simply" dig through the human genetics literature and look for associations between genetic variants and human traits. They catalog these and write brief narrative descriptions of them:

> rs6457617 has been reported in a large study to be associated with rheumatoid arthritis. This SNP is reported to be the most statistically significant of many SNPs similarly located in the MHC region. The risk allele (oriented to the dbSNP entry) is (T); the odds ratio associated with heterozygotes is 2.36 (CI 1.97–2.84), and for homozygotes, 5.21 (CI 4.31–6.30).[16]

What the hell does this mean? To start with, "rs6457617" is the SNP number; that is, it is the unique identifier of a particular variant in human DNA ("rs" stands for "Reference SNP," one that has been validated and mapped to a particular place in the genome). Now, recall the DNA alphabet: A, G, C, and T. Our genomes are the 3 billion A's, G's, C's, and T's we get from our mothers and the 3 billion we get from our fathers. Of the thousands of people who have been screened for this particular SNP associated with rheumatoid arthritis, virtually everyone on earth is one of the following: CC, CT, TC, or TT. People who inherited a C allele at this SNP from each parent are at average risk for developing rheumatoid arthritis. People who inherited a T from one parent and a C from the other at this position are 2.36 times more likely to develop RA than average. People who inherited a T from both parents (as I did) are 5.21 times more likely to develop RA.

Okay, but what does that mean in *absolute* terms? We don't know with 100 percent accuracy, but after typing me for this SNP and five others associated with arthritis, Navigenics told me that my lifetime risk of developing RA was 2.8 percent, or a little less than twice the average. The first big caveat: from twin studies, we know that only slightly more than half of the risk for RA is inherited; the rest is likely due to environmental factors, which Navigenics has not measured (nor, as far as I know, have any of the other commercial or noncommercial genome scanners, probably because no one knows exactly what they are).[17] The second big caveat: there's no reason to think that scientists won't find another dozen SNPs in the human genome that contribute to RA susceptibility. The model for risk prediction in RA will probably look much different in a few years (if not months) and it will probably be much more complicated.

But for Mike Cariaso and Greg Lennon, that wasn't the point. Over a long lunch at a cheap French restaurant in a nondescript part of Bethesda, Maryland, near the hotel where NIH likes to hold meetings, Greg recounted the genesis of SNPedia.[18] He took me back to 2005–2007, a simpler time when there were no direct-to-consumer genomics companies or gargantuan databases brimming over with information on human genomic variation. This state of affairs didn't sit right with Lennon, a handsome man in his early fifties with thinning gray hair and bright blue eyes. He spoke in relaxed, measured tones, although one sensed impatience just below the surface. In the early 1990s he had finished his postdoc with übergeneticist Hans Lehrach at the Imperial Cancer Research Fund in London and had followed that with a successful career as a biotech scientist and executive. He and Cariaso met when both were working at Larence Livermore in northern California. Cariaso then followed Lennon to Gene Logic, one of the first companies to take seriously the idea that gene expression—to what extent certain sets of genes were active in particular cells and tissues—could be used to identify drug targets. Thus, for example, white blood cells express high levels of genes that code for infection-fighting proteins; neurons express high levels of genes that code for neurotransmitters such as dopamine; and so on. To a large extent, different cell types can be defined by the genes they do or don't express. (Alas, thus far this has not led to much in the way of drugs.)

The lingua franca for measuring gene expression in the late 1990s and early 2000s was the microarray: tiny spots of DNA fixed to a solid surface such as a glass microscope slide or nylon membrane. A microarray typically contains thousands of genes. To survey the expression of those genes in a cell or tissue with a microarray, one would prepare fluorescently labeled RNA (the intermediate coded for by DNA that usually goes on to code for protein) from that cell type. All of those bits of RNA serve as probes: they find their complementary DNA partners and stick to them like molecular Velcro. When they find their match, they fluoresce. The strength of the fluorescent signal that lights up at the spot in the array representing each individual gene provides a snapshot of how active that gene is in the sample. Genes that are especially active or inactive in diseased cells and tissues are potential drug targets.

But by 2000, microarrays were beginning to be used for purposes beyond gene expression. Among the new applications was SNP detection, and this was even easier than gene expression. Instead of a curve measuring the extent to which a gene was expressed, genotyping was binary: In any given individual, was a particular DNA variant present or absent? And if present, was there one copy or two? By doing case-control studies on hundreds or thousands of people, say, half with a certain disease and half without, and by finding SNPs that were more frequently found in those *with* the disease, SNP scans could be used to identify disease susceptibility genes.

The leap Cariaso and Lennon made was to take those findings and begin to apply them to individuals. Because once those genome-wide association studies (GWAS) were done on one or more populations, then hypothetically anyone could examine some fraction of her complement of SNPs and see whether she carried SNPs that raised or lowered her disease risks or otherwise contributed to her traits. In 2007 this was still an expensive, labor-intensive, and high-risk proposition for two guys in their basements. So why even bother?

"I was frustrated," Greg Lennon told me. "I could go into any restaurant and ask, 'Has anyone here benefited from the Human Genome Project? Do you know *anything* about your own genetics? Do your doctors know anything about it?' The answer in general would be—and for the most part, still is—a resounding no. Not even at the level of cocktail banter. I spent my career in an area of science that I felt and continue to feel is very promising. Yet it doesn't seem to matter. It hasn't affected anybody. So I began to ask myself, 'Have I just wasted my time?' "[19]

After talking about it for more than a year, Lennon and Cariaso decided to go native. From Gene Logic's deep and abiding work on gene expression, the two were quite familiar with Affymetrix GeneChips, the dominant microarray platform at the time.[20] They had both done some genotyping. So how hard could it be to run some Affymetrix chips on themselves and have a look at their own genomes?

Arguably the pair's most difficult hurdle turned out to be getting DNA out of their own bodies. Spit kits and spit parties were still many months away. Blood was easier and cheaper to process. But there was a problem. "When you're a random individual wandering the streets," Lennon said, "no one really wants to collect blood from you."[21] Cariaso stopped in a fire station in suburban Maryland and found that paramedics have a lot of time on their hands between calls. He chatted them up, got them interested in what he was doing, and soon had his sleeves rolled up. And that would have been that, except . . . "Sitting in the back of an ambulance with the needle in my arm, the station alarms went off."[22] Another possibility was the Red Cross: Lennon told me that if you had "the right attitude," then you could get some of your own blood to go. He wound up getting his own blood sample from a general practitioner who had a soft spot for human genetics, despite the admission that he remembered almost nothing from his cursory medical school training in the subject.[23]

Lennon and Cariaso found a contract molecular biology lab to isolate DNA from their white blood cells and to run the latest and greatest Affy chip (five hundred thousand markers). According to Lennon, that's when the fun began. "We got that data back, and for all of our brilliance, we just stared at those huge files going, 'Now what?' "[24]

They were convinced that at some point during the thirteen-year, $2.7 billion effort to map the human genome, surely *someone* had taken the initiative and developed a database that systematically linked variation across the genome to human phenotypes. There was Online Mendelian Inheritance in Man,[25] an incredibly useful tool developed by the late Victor McKusick, the father of clinical and medical genetics. The catalog began as a hardcover book, *Mendelian Inheritance in Man*, in 1966.[26] But even though it now lived online, OMIM was a text-based catalog, not a digital one, and it mostly contained diseases and phenotypes caused by rare changes in single genes: if a doctor in Saudi Arabia observed a child with widely spaced eyes and elevated enzyme levels in 1968 and published a case report, McKusick would make a note of it. The catalog was and is remarkably comprehensive: an astounding collection of our species' variation.[27] That said, I often found reading OMIM to be annoying: helpful and fascinating case reports and research studies were amassed under gene and/or disease headings and subdivided ("clinical features," "animal models," "pathogenesis," etc.), but without any real narrative flow. At its best, it was like an Audubon field guide for clinicians—a terrific, handy reference. At its worst, it could be a painful slog for anyone interested in a high-level view of any particular genetic disease. I would be terribly upset if it didn't exist, but as the great Irish writer Roddy Doyle said of *Ulysses*, it might benefit from a little more editing.[28]

The public database of genetic variants, dbSNP, began in 1999. In 2002 it contained 1.3 million unique, validated human variants. In early 2010 it had 9.5 million.[29] But unlike OMIM, dbSNP had no intrinsic clinical content, and until recently it didn't "talk" to OMIM. "I respect McKusick and the way he put OMIM together," said Lennon. "But that doesn't mean it's kept up with the times. It doesn't help you annotate your genome. The vast majority of the information is effectively anecdotal. There's nothing wrong with that. But you can't actually have software work with OMIM. At least dbSNP had the nomenclature part of it roughly right."[30]

Disappointed that the billions spent on the Human Genome Project had not resulted in a resource linking genotype to phenotype, Lennon and Cariaso were at a loss. Despite their ambition, the idea that the two of them could take all of the world's human genetics literature and turn it into useful information for the benefit of a dozen or a thousand people was laughable. The idea that *any* number of people could do it was debatable. Lennon and Cariaso could pour the foundation, but other people would have to finish the floors and furnish the house. It would take a village . . . or at least, something like Wikipedia. And that was the approach they took: the two put other people's SNP data up on the site (they never got around to posting their own) and let the world have at it.

"We faced and still face the exact same questions that Wikipedia faces," said Lennon. "How do you control quality? Is the information credible? Fine—those are fair things to ask. Why not be skeptical about anything you read? On the other hand, we have a huge advantage over Wikipedia because we're not trying to cover everything from the Israeli-Palestinian conflict to Britney Spears. And we don't get a lot of flame wars on the site. Genetics is usually pretty boring."[31]

Nor was SNPedia trying to turn a profit like the personal genomics companies. It has been free and open-source from the beginning. It does not perform experiments and it does not offer its own interpretations. It simply mines the literature and reports conclusions drawn by others.

After Marco Island, I got my Affy500 SNP data—five hundred thousand markers—from George and sent it on to Mike. All three of us were gung ho, perhaps a little too much so. George was keen to send it to me because the PGP was still trying to figure out the interpretation part of the equation and he thought maybe I could do some of the legwork to see what was out there. Mike was keen to get his paws on PGP data and anything else people were willing to share—more data points for SNPedia could only help the site to grow and be taken more seriously. And I was jazzed because I would finally get a glimpse of my own genome. Mostly jazzed anyway:

———- Original Message ——

From: Misha Angrist
To: cariaso
Sent: Saturday, March 1, 2008 7:23:57 PM
Subject: Fw: Drum roll please

Hi Mike:
    If you're still willing to run a report, here are my SNP data. You are free to add me to the roster of public genomes, though, I don't know, maybe I should take a day or two to

look the data over first? I can't imagine redacting anything. Anyway, maybe the thing to do is to run the report and we can talk. Would there be anything to gain by sending it to Scheidecker or anyone else? I guess if they care they can get it from SNPedia.

Thanks again,

Misha

From: cariaso
To: Misha Angrist
Sent: Saturday, March 1, 2008 8:13 PM
Subject: Fw: Drum roll please

ok, the report is being generated. I'll probably send it to you later this evening. You can review it before taking any action on the rest of this email. If you decide you'd like to share it I have to point out. I have not shared my report with anyone, so I can relate to your concerns. But if you're having second thoughts about sharing your SNPedia data, you should know you're also one of the PGP-10. It's not a question of if this will come out, only when.

That night Ann and I went out to a benefit dinner. We were going through a rough patch. I was depressed, emotional, and on edge. I was living inside my head, as is my wont, and at that time it felt like an especially ugly place to be. I was feeling paranoid, constantly trying to parse what people said to me, taking every perceived negative, no matter how slight, to heart, and convinced that anything positive that came my way couldn't possibly be sincere. There's an old Loudon Wainwright III song that goes, "I wonder why you love me, baby / I hardly love myself at all."[32] This was my theme song. I was sleeping badly. I was angry at Ann, angry at my kids, angry at my therapist, and most of all, furious with myself for being forty-three years old and such a miserable fuckup. Self-loathing was my specialty. As we got out of the car, I told Ann I would be the fourth public genome on SNPedia. She rolled her eyes and asked why. Why did I always have to act so impulsively? I said I thought we had gone over this: that I would be public, that this would be about demystifying DNA for everyone, that we were not our genomes. She didn't want to talk about it anymore. "I don't like how we've left it," I said of our discussion. "Sometimes you won't," she said.

When we got home I was still in a cantankerous mood; Ann fled to bed. I checked my email and saw that Mike had run my Promethease* report. With some trepidation I opened the file. As I scrolled through it, I knew I should be clinical, objective. I had been a genetic counselor, for God's sake. I knew enough about statistics to know that having a risk ratio go from 1 to 1.6 or even 2 meant my absolute risk had only risen from, say, 1 in 10,000 to 2 in 10,000. But still, it was unsettling. And what about conditions where there were multiple alleles in multiple genes—were the risks additive, multiplicative, or something else? Could one bad version of a multiple sclerosis gene undo the health effects of four good versions of other MS susceptibility genes? Was I really likely to get rheumatoid arthritis? Is this why my fingers were always so achy in the morning?

I was drunk, tired, and now fairly certain I was going to succumb to autoimmune disease in the coming months. I began bombarding Mike with questions:

> *What about the SNPs that say nothing specific about my risk? Can I assume it is "normal"? Or just unknown? And what about conditions where there are multiple SNPs—are the risks additive, multiplicative, or epistatic? Do I just have to go digging in PubMed?*

PubMed. Start on the rs# page, digg a little bit. Even if you don't 100 percent figure it out, if you leave notes about

---

* Prometheus was the Greek Titan who stole fire from Zeus and gave it to man. The suffix -*ase* is typically attached to names for enzymes. Proteases, for example, break down proteins. Lactases break down lactose.

what you think you found where, someone else may be able
to clarify.

*What about the ones that state a risk but for which there are no
text links?*

give me an example.

What are the #s in the far left column in pink?

You're not even trying. Click on help at the top of the report.[33]

I wasn't even *trying*? Ouch. Even if that were true, in the genetic counseling handbook, saying that to someone—even someone as irrational as
me who ought to have known better—was a huge no-no. How dare he!
With the click of a mouse I had gone from an ordinary neurotic/depressive
who didn't believe in genetic determinism to a quivering mess who "knew"
himself to be a ticking time bomb. The person most likely to talk me off
the ledge was both fed up with me and, incidentally, asleep down the hall.
Meanwhile the guy on the other end of the screen had had enough, too.
My blasé attitude was being tested. We were at the first of many "what in
the name of Gregor Mendel does this all mean?" moments. For all of the
hundreds of studies that had been done on tens of thousands of research
subjects with and without various diseases, the truth was, nobody knew
an individual's risk for complex diseases with certainty.

It was 1 A.M. I could hear all of the naysayers at the *New England Journal of Medicine* laughing at my mini-eruption of angst. *You were so sure that
this information was harmless. How does it taste* now, *Genomeboy?*

I called Dietrich Stephan. "I think I'm ready to take you up on your offer of
a freebie," I said. "That's great," he said. "I'll get a kit in the mail to you."[34]

A few days later I received a "spit kit," which was basically a tube I was
to fill with saliva. *For best results, collect your sample just before eating a meal and*

when you are in good health. . . . *To make more saliva, close your mouth and wiggle your tongue or rub your cheeks.* Although I was someone who still drooled when he played the guitar, I found that this took a bit of doing. After a few minutes I reached the fill line. I put the plastic tube in the Styrofoam box and sent it off to Affymetrix, the California DNA-chip company that was processing Navigenics' samples.

In a couple of weeks I received an email that my "HealthCompass Profile" was ready. I didn't rush to look at it. After my semidark night of the soul with SNPedia I didn't want to barrel through the process this time. After work that day I dithered around, ate dinner, put the kids to bed. When my inbox was finally empty and I had sedated myself with a full thirty minutes of *Puppy Bowl IV* on Animal Planet, I figured it was time. I logged on. *Welcome! Your results are ready!* Genetic counselor Elissa Levin's smiling face was there to reassure me. The conditions were divided by "estimated lifetime risk": < 1 percent, 1 percent–10 percent, 10 percent–25 percent, 25 percent–50 percent, and > 50 percent. Each condition had its own box containing the disease name, my results, and the average lifetime risk for males. If the box was orange, I might want to pay closer attention because that meant either my risk was 20 percent or more above the population average, or my lifetime risk was greater than 25 percent. I was beginning to feel nervous despite myself. My palms were sweaty.

Alzheimer's was first. This was APOE, the gene Watson and Pinker didn't want to know. If I had one or two copies of APOE4, I'd be at higher risk. At the time, Navigenics did not test directly for the APOE gene, but rather a marker near it. This was partly for technical reasons and partly because the APOE gene was patented (we'll revisit this notion). But the marker near it was a nearly perfect proxy for APOE. So did I have one copy of APOE4? Two? I clicked . . . zero. Navigenics said the average lifetime risk of Alzheimer's for males was 9 percent, which sounded about right; my risk, given my APOE status, was 5 percent. I felt a palpable sense of relief and a little guilty for allowing myself to feel that. But the fact that I had zero copies of APOE4 meant each of my parents could have no more than one copy. I wanted to call them and tell them, even though that would have been a deterministic thing to do.

Celiac disease: average risk was 0.06 percent; mine was 0.02 percent. Woo-hoo. Bring on the gluten.

Colon cancer: average risk was 6 percent; mine was 5 percent.

Crohn's disease: average risk 0.58 percent; mine was 0.37 percent. This was based on my genotype at nine different genes.

Type 2 diabetes: average risk was 25 percent; mine was 32 percent. Whoops—time to lose weight, exercise, and maybe get a glucose-tolerance test. As if I didn't know.

Glaucoma: average risk was 1.1 percent; mine was 3.4 percent.

Graves' disease: Population risk = 0.55 percent; me = 0.93 percent.

Myocardial infarction (heart attack): Population risk = 42 percent; me = 38 percent. Wow . . . it was staggering to think that the average lifetime myocardial infarction risk in males is 42 percent. My paternal grandfather died of a heart attack at age fifty; my dad had one at age sixty, as did my mom's mother. This was another obvious message that I needed to exercise, watch my diet, control my cholesterol, etc. What a shame: I had so wanted to take up smoking.

Lupus: Population risk = 0.03 percent; me = 0.01 percent.

Age-related macular degeneration: Population risk = 3.1 percent; me = 1.1 percent. My grandmother had it. I wondered what that did to my risk.

Multiple sclerosis: Population risk = 0.30 percent; me = 0.17 percent. But this was based on just three markers. MS risk was mediated by dozens of them.

Obesity: Population risk = 34 percent; me = 36 percent. Does this genome make me look fat?

Rheumatoid arthritis: Population risk = 1.1 percent; me = 2.8 percent.

Restless legs syndrome: Population risk = 4 percent; me = 4.3 percent.

Psoriasis: Population risk = 4 percent; me = 2.5 percent.

Prostate cancer: Population risk = 17 percent; me = 12 percent.

These were interesting numbers, but they didn't rock my world. By now I had gotten a grip. My biggest concern was about metabolic syndrome, basically a group of risk factors for heart disease: obesity, bad lipid numbers, high blood pressure, insulin resistance, etc. Two years earlier I had been in an exercise study because I was already known to be at high risk for this stuff and I had a modest history of diabetes in my family.

Now I could hear the geneticists saying, "We told you so," for another reason. Navigenics' service was fast, efficient, thorough, and impeccably designed and presented. Its explanations of genetics were simple and clear without being condescending. But there's no way I would have spent $2,500 (the cost in 2008) on it. The problem of course was not that the information was useless; it was that genes told only part of the story of one's health and so far we were only a few chapters into a very long and complicated tale. Thus, until we started linking environment (as a child did I have lead paint in my house?), phenotype (how much did I weigh?), and lifestyle (what did I have for breakfast?) with personal genomic profiles, it was not realistic to expect that the latter would tell us things that would save—or even dramatically improve—our lives.

That said, in the wake of my genome scans, I began to see my iconoclastic doctor more regularly, I went back to the gym, and I became an aspiring vegetarian for a while (Chinese pork buns were still my weakness). Eventually, with the help of my kids, I got into the Wii Fit. Is this because I worshiped in the Temple of the Gene? No, it's because the genetic evidence reinforced the story told by my family history: I was at risk for heart disease. Getting my genome done and collecting my medical records for the PGP was a convenient excuse to begin to get my cardiovascular act together.

To its credit, Navigenics constantly updated its results based on the latest science. A year after I received my initial analysis, my risks had changed for many of the conditions, albeit not very dramatically. And where there had been seventeen diseases before, by 2010 there were twenty-seven. I learned that my risk for osteoarthritis, for example, was a whopping 56 percent based on two genes. I clicked on the button that read "What you can do." Early detection was a good idea, Navigenics said, in order to distinguish joint pain from other forms of arthritis such as rheumatoid arthritis. X-rays, MRIs, and a variety of blood tests might be in order. To prevent it I could lose weight—this seemed to make a big difference as to whether one developed osteoarthritis in one's knees. Other things to consider, the site said: rubber-soled shoes, no high heels, regular exercise.

This was largely common sense, of course. Osteoarthritis, it seemed to me, was simply collateral damage that came with getting old. One could understand why a customer might peruse her results and recommendations and say "meh." Meanwhile some geneticists and ethicists had started to complain that these tests would lead to a "raid on the medical commons."[35] In other words, with their genomic results in hand, otherwise healthy folks would begin availing themselves of every medical test they could. The high-tech genome scan would demand a series of high-tech fixes. Genome scans would be the new CT scans—sexy but imperfect and usually unnecessary. And this would place an additional burden on the health-care system and drive up skyrocketing insurance premiums even further.

I supposed that could be true. But I wasn't sure that the answer was for people to willfully ignore their genomes. Hadn't we scientists spent years telling people how fantastic this stuff was all gonna be? Didn't we justify our ongoing suck at the federal teat by promising diagnostics and cures? Didn't we want to encourage people to take responsibility for their health and to embrace preventive measures? Hypochondria was a risk, no question. But the price of genomic information was falling every day: indeed, in the summer of 2009 Navigenics lowered its price 60 percent to $999.[36] I still believed that it was incumbent upon us—the genetics community—to view this as an opportunity, not for more CT scans, but to talk about whether people really needed CT scans, what sorts of healthy behaviors they could manage on their own, and how their genomic data could play a part in that.

For decades, medical genetics has been criticized as a field akin to bird-watching, whose credo is "diagnose and adios." In 2010, our ability to treat strongly genetic diseases was fairly pathetic. But even in those many instances where we could do little or nothing to mitigate our genetic lot in life, a few scientists and physicians were starting to ask whether we might want to know what our risks were anyway.

8

Gettysburg to Gutenberg

In the late 1990s, Robert Green, a neurologist at Boston University, asked his geneticist friends a simple question: Should we start offering genetic testing that would let people know their genotypes for the APOE gene, that is, tell them if they were more likely to get Alzheimer's? The geneticists were mostly horrified. Not only was there nothing one could do to delay the onset of Alzheimer's, they told him, but the average person would almost certainly misunderstand his/her risk, come to serious psychological harm, and maybe even commit suicide. This was crazy talk.[1]

Green is a good-looking, bespectacled man—a dead ringer for Pittsburgh Penguins coach Dan Bylsma; as it happens, he has the coach's same understated manner. His voice is a pleasing tenor: while he was on the faculty at Emory University, his performances as part of a chorus that sang with the Atlanta Symphony won five Grammys. Unlike the disheveled basic scientists he often hangs around with, Green favors a jacket and tie. He is a self-described "clinical trialist" who for many years studied prospective—and mostly disappointing—Alzheimer's treatments.

But perhaps, he thought, he could do some good by providing Alzheimer families with *information*. Not only did he reject the standard "this is toxic knowledge" stance of the geneticists,[2] but the idea of disclosing risk and studying its effects actually appealed to him as both a clinician and a scientist. So he did what he always did: played by the rules and set up a randomized clinical trial. One group of people with a family history of Alzheimer's would receive its genotype information; the other would not. The group that did could get 1) good news (no copies of the APOE4 allele and therefore a 9 percent lifetime risk of Alzheimer's); 2) moderately bad news (one copy of APOE4 and a 29 percent lifetime risk); or 3) really bad news (two copies of APOE4 and a greater than 50 percent lifetime risk). What Green's team found was that those getting bad news would indeed get bummed out (just as I did when I learned about my phantom multiple sclerosis) . . . but not for long. And virtually no one regretted learning his or her risk.[3]

It turns out this is the norm, whether the devastating disease is Alzheimer's, breast cancer, ovarian cancer, or Huntington's disease. "There are dozens of articles and they all seem to show the same thing," Green said. "In the short term people may become upset by positive results [showing they are at high risk], but they cope quite well. And when people learn they have negative results, they are relieved and that relief persists."[4]

Until recently, most of the genetics and bioethics community was having none of it. As we sat in the living room of his stately house in Wellesley, Massachusetts, Green wrapped himself in a blanket on the couch and told me that over the years his studies had been called irresponsible, unethical, foolish, and unrealistic. "They said our careers would be over."[5] Now his work was held up as a template for how to disclose genetic and genomic information. Green eventually joined the PGP team as a counterweight to George's rebellious tendencies and flights of fancy: an even-keeled, NIH-savvy physician who understood the ins and outs of delicate human subjects research on genetics. "Bob was brought in specifically to keep me on a leash," said George with an impish smile, implicitly acknowledging that that would be no mean feat.[6]

At least two personal genomics companies included APOE as part of their SNP-based scans (Green consulted with them pro bono).[7] As genes go, APOE is arguably the most powerful one yet discovered that contributes to complex traits mediated by multiple genes and the environment. This is probably why both Jim Watson and PGP-10 participant Steven Pinker declined to learn their APOE genotypes (though, as we saw in chapter 7, successfully *not* knowing was not always so easy to accomplish).

Time was short. The National Human Genome Research Institute (NHGRI) folks were about to pay a visit to the Church lab in anticipation of George submitting a renewal for his big grant, one of ten such genome grants in the country. The site visit would not be a make-or-break; indeed, it would not be evaluative at all, only advisory. The NIH wanted to know what progress the Molecular and Genomic Imaging Center had made over the last five years and what George's team was planning for the next five. The idea was to maximize the grantee's chances for success going forward. George thought the stakes were fairly low and refused to get worked up about it the way Bob Green and Jason Bobe had done. "The only tricky thing," George conceded, "will be getting the PGP right."[8]

"Tricky" was an understatement. From the moment the NIH notified George that it would fund every aspect of the imaging center *except* the PGP, rhetoric and accusations flew back and forth between the Church camp and NHGRI. From what I could glean from these exchanges, the NIH saw George as a brilliant scientist, a world-class biologist, and a sharp and incisive mind that had always rewarded the public's investment in him. Perhaps even more important, he was a card-carrying inventor and precocious technology developer who had a knack for seeing the future. It would be stupid not to make him a centerpiece of the institute's efforts to realize the thousand-dollar genome. "We ignore George Church at our peril," a powerful genome scientist once said. On the other hand, there were parts of the Churchian future that the NIH simply could not stomach. His views about genetic privacy, for example, which he began to spell out in the original 2003 Molecular and Genomic Imaging Center proposal:

> The core question is: How may the gathering of increasing amounts of genetic information be made compatible with ethical and legal requirements for privacy? Anything approaching a comprehensive genotype or phenotype (including molecular phenotypes) ultimately reveals subjects' identity in our increasingly wired world as surely as conventional identifiers like name and social security number. . . . This raises numerous specific questions:
>
> - Are current informed consent practices sufficient to give human subjects adequate understanding of the potential that their identity may be discernible in large genetic data sets . . . ?
> - Is enough protection afforded by allowing researchers open access to such data sets so long as they agree not to take the analytical steps that would link these data to a specific person . . . ?
> - Is there a kind and level of genetic information for which it would be virtually impossible for a researcher . . . to link it with a specific person?[9]

For George, the answer to each of these questions was a resounding no. DNA was the ultimate digital identifier, after all: a Social Security number was nine digits, while a genome was 3 billion. George believed the NIH was balking at paying for the project (despite having approved every other aspect of his $10 million genome technology grant) because he refused to do it under the ethical paradigm set forth by the agency—that is, one in which subjects give informed consent and in return are promised, more or less, privacy and confidentiality.

And what about the long list of privacy breaches involving health and genetic information?[10] To NIH, these might have been noteworthy anecdotes that raised potentially provocative questions about the privacy of genomic data. Such questions were certainly worthy of study. But their answers were in no way self-evident. It was one thing for the Church team to come up with faster ways to sequence human DNA or better ways to interpret what those sequences meant. It was quite another to start sequencing healthy humans, sharing every detail of their genomes with them, and not even *trying* to keep their identities secret.

In early 2007, Jeff Schloss, the National Human Genome Research Institute's point person on sequencing technology grants, reluctantly granted me a brief telephone interview. His main point was that George had not made the case, either in his 2003 grant application or in three years' worth of subsequent correspondence, as to why the PGP should be carried out under different rules of informed consent. "We need hypotheses," Schloss said. "We need to know what scientific ideas he's going to be testing."[11]

"We're doing hypothesis *generation*, not hypothesis *testing*," countered George. "What hypothesis were we testing when we sequenced the first human genome? When he sold the Human Genome Project to Congress, Jim Watson didn't say we were going to test hypotheses, he said we were going to cure cancer."[12]

A few weeks later I tagged along to a meeting of the various ethics centers funded by NHGRI. There an NHGRI administrator gave a presentation outlining the institute's plan to sequence five to ten humans in the next one to three years. It was never clear to me what NIH hoped to gain from this, as there was no discussion of scale-up or data analysis. Multiple companies had, like George, already begun to sequence humans to demonstrate the power of their technologies. But NIH itself was not in the technology development business—it gave money to academic scientists and start-ups to do that. And clearly NHGRI was not excited about publicly identified genomes. During the Q&A someone asked about the PGP-10 and how it was different from what NIH was proposing. An administrator said that

George insisted on sequencing highly educated geneticists and seemed to imply that this was elitist, celebrity genomics. The interlocutor, an ethicist familiar with the PGP, said she thought that the education requirement was something that had been demanded by Harvard. The administrator said no and insisted it was George's idea. I had interviewed Rabbi Terry Bard, a member of the Harvard Institutional Review Board that oversaw the human subjects aspects of the PGP. He told me that having the PGP-10 be highly credentialed was indeed the IRB's stipulation, not George's.[13] Yet in the coming months I would hear the same charge leveled at George from several people affiliated with NHGRI.

I paid a visit to the fortress itself. On the NIH website there was a warning about "new security procedures." I showed up early, which was a good thing. I drove into the car inspection line and queued up, and a succession of orange-vested people approached my window. The first brandished a metal detector wand, which she waved over my steering wheel. The next collected my driver's license. The third placed a sheet of paper on my dashboard saying that my car had been inspected and would I please open the trunk. The fourth asked me to step out of the vehicle and walk through a metal detector.

I should say that, through the entire process, there was not an ounce of DMV- or airport-security-style surliness. Judging by my experience at the entrance gate, NIH was populated by Shiny Happy Civil Servants. They preempted one's inclination to get mad at them. They had to do this all day in the sweltering heat—so if they weren't irritated, how could you be? I returned to my car and a smiling orange vest came back with my license and a badge with my name on it. Then I sat and waited. And waited. And waited some more.

Eventually I was waved through. I found Building 49 and waited for Les Biesecker to come downstairs and guide me over the final security hurdle. Biesecker ran the Genetic Disease Research Branch at NHGRI. He was tall with short gray hair and glasses on the end of his nose. Biesecker was in charge of ClinSeq, which its website described as an attempt to "pilot large-scale medical sequencing in a clinical research setting."[14] He told me that he and his colleagues wanted to bring genomics to bear on medical problems. The idea was to find a population with a medical condition and sequence a couple of hundred genes in each patient where there's likely to be a "hit": a mutation or mutations that are important in that disease. The National Heart, Lung and Blood Institute bit on this idea: cardiovascular disease would therefore be ClinSeq's initial focus.[15]

Like the PGP, ClinSeq would return results to participants. But unlike the PGP, it would return them only if they were clinically relevant. This meant 1) known genetic variants that led to known clinical manifestations; 2) variants that appeared to be severe because they stopped a complete protein from being made, they inserted/deleted sequence, or they drastically rearranged the gene; and 3) more ambiguous changes that altered amino acids within a protein but whose effects were not known. "Those will be tough," said Biesecker of the last category. "The good thing is that for cardiovascular disease, there are interventions, so if there is bad news then there are potentially things that can be done [in those cases]."[16] For each gene analyzed then, ClinSeq staffers would have to investigate the variants they found and decide if a case could be made for returning results from that gene to subjects.[17]

"Why not return all of the variants or even all of the sequence data back to participants who want it?" I asked.

"I'm not interested in the genome as entertainment," he said.[18]

When I got back to my car, I noticed a sign on the top floor of the parking garage. ATTENTION: OVERHEAD LEAKS OR BIRD DROPPINGS MAY STAIN CAR FINISHES. PARK AT YOUR OWN RISK.

In my twenty-five years of parking I had never seen such a sign. I imagined it was dreamt up by a government lawyer somewhere as some form of preemptive legal strike against angry drivers. But maybe that was too cynical—maybe it was just full disclosure: "Here are the possible consequences of parking here. If you park here then you consent to the possibility of bird shit on your car." Okay, good to know—two cheers for NIH. But bird droppings or unspecified overhead leaks were not in the same league as genomic information. Given the gauntlet one had to traverse just to pay a visit to NIH, it was easy to understand why the institutes appeared to have zero interest in health information altruism or open consent. Openness implied risk. In acknowledging how little we know about what any of the information written in our genomes means, Amy McGuire once said we must learn to consent to uncertainty.[19] But bird poop notwithstanding, that was a lesson our government seemed unlikely to assimilate anytime soon. Uncertainty was still Kryptonite.

While fighting for NIH acceptance, George continued to pursue private funding sources. Although George would not discuss it, others familiar with the PGP told me that at one point Google was poised to infuse tens of millions of dollars into the PGP to help develop sequencing technology and use the project as a showcase for Google Health, the company's new Web-based platform for central storage and management of all of one's health information. Before pulling the trigger, however, Google higher-ups wanted to solicit the opinions of other thought leaders. What effect would public disclosure of DNA and trait data have on the public's view of a company whose unofficial motto was "Don't Be Evil"? The PGP brain trust had a sit-down with NHGRI leadership, a few bioethicists, and a couple of senior Googlers. According to a PGP affiliate who was there, "It was clear we were walking into a hornets' nest." After the meeting, sources told me, NHGRI leadership followed up with an email to someone on Google's board of directors saying that if Google went ahead with the PGP it could set personal genomics back twenty years. At that point the collaboration was all but dead.

George was distraught. He had imagined Google Genomes taking its place alongside Google Books, Google News, and Google Maps. Admittedly, the PGP still had a million bucks from the company. But that is all it would get for the foreseeable future.

"For one brief shining moment," said Ting, "it was Camelot. George had gotten to be good friends with the Googlers. It had become kind of an extended lab situation. So when it ended it was hard."[20]

All of this pushback made the inclusion of the PGP in George's grant renewal proposal a piping hot potato. George still desperately wanted NIH's approval for the project, but after four years of supplication, argumentation, butting of heads, and politicking, he was no longer sure how to get it.

Enter Robert Green. In preparation for the site visit, he began meeting with George and Jason Bobe several times a week. If this proposal were to succeed, he told George, the PGP would have to be scaled down by at least two orders of magnitude. He laid out some ground rules:

- No use of the phrase "100,000 people" (500 max, or better yet, 250)
- No discussion of asking participants to pay their own way
- No insistence on a total lack of privacy for participants
- Come up with a very modest plan to collect trait information
- Rehearse your PowerPoint presentations ad nauseam

"If George's goal is to get NIH funding for a portion of the PGP," Green said, "there really isn't any choice. You have to play this game. One has to abide by the standards of scientific proposals that have become the norm at NIH."[21]

Under Green's tutelage, the Church team fell into line. Everyone dutifully made slides and practiced their talks while Green sat in the audience and played the role of hostile questioner. But the dry runs seemingly paid off: the new, scaled-down incarnation of the proposal—now with the even more unwieldy moniker, "The Human Gene/Environment/Trait Technology Center" (HGETTC)—was ready. Two days prior to the site visit, Jason was giddy. "This is going to be like Gettysburg," he said, invoking a George metaphor.

"You mean a bloodbath?" I teased him.

"No," he laughed. "A turning point!"[22]

. . .

Gail Henderson had never heard of George Church. And even after reviewing the materials that had been sent to her, she still didn't know quite
what to make of the Personal Genome Project. But since she was the member of the site-visit team responsible for ethical, legal, and social-issues
(ELSI), all eyes were on her. Henderson was professor of social medicine
at the University of North Carolina Chapel Hill, about ten miles from
where I work. "Durham is a full day's journey from Chapel Hill," she told
me half jokingly over beers, referring (I think) to both the cultural divide
between Duke and UNC and the increasingly snarled traffic on the highway connecting the two cities.[23]

Henderson may have been new to the PGP, but you would have been
hard-pressed to find anyone who knew more about ethics and biomedical research. She had long studied how scientists, patients, and ordinary
people perceive genetic research and genetic medicine. And now, as director of UNC's Center for Genomics and Society (an ethics center funded
by NHGRI), she had turned her attention to ethical and regulatory issues
posed by new genomic technologies and the big science that came with
them. The PGP was right in her wheelhouse. "Our center's theme is about
exactly this: What are the ethical, legal, and social issues when you start
to scale up?"[24]

The night before the site visit, Henderson caught a plane from Ra-
leigh to Boston. When she arrived she felt tired, out of breath. Her heart
raced whenever she stood up. She thought she was exhausted from traveling and stress: after the site visit she had a meeting in D.C. and from there
was scheduled to go to China, where she was to continue her studies of
the Chinese health-care system. She reached the Best Western near Harvard Medical School and went straight to the restaurant—maybe she just
needed to eat something. She sat down . . . and promptly passed out and
fell to the floor. When she came to, two strangers were kneeling over her.
They called an ambulance, which didn't take long. "If you're gonna pass

out," Henderson said, "the place to do it is directly across from Brigham and Women's and Beth Israel Deaconess hospitals, right?"[25]

She was experiencing a slow but massive gastrointestinal bleed, though she did not fully apprehend its severity at the time. "I had this idea that this was nothing," she said. "I talked to the NIH and said I'm *sure* I'll be out of here soon; put me on the agenda for tomorrow afternoon."[26]

Once word spread that Henderson was in the hospital, there was concern on both sides, not only for her health, but for what this would mean for the site visit. The Broad Institute's Chad Nusbaum, a member of the NHGRI team whose job was to evaluate the sequencing technology, was worried that he would be called upon to have "intelligent opinions on ELSI matters."[27] Jeantine Lunshof, the philosopher/bioethicist who had helped George articulate the rationale for open consent, fretted over the prospect of not having a fellow ethicist in the room.[28] And Jason Bobe feared the worst. "I thought this was a sign that NHGRI wanted to bump this part of the proposal altogether."[29]

Robert Green sensed the apprehension and decided to act. "It's always good to meet people face-to-face," he said. "The value of the site visit was that we were going to be able to talk person to person. It was not a calculation, but an instinct. I went to her hospital room and found her fully lucid and ready to talk. We started to go through the slides, but she stopped me and said, 'Why don't you just tell me about this?' And that turned into a conversation that went on for over an hour."[30]

By the time Henderson was piped into the meeting by phone, most of her reservations about the PGP seemed to have been assuaged. "[When she addressed the meeting], she ended on this enormously positive note," said George. "The impression I got from NHGRI was that they thought we were really doing our homework on the ELSI."[31]

What about everything else George wanted to accomplish? What about the RNA, immunogenetics, the microbiome—i.e., the collective genomes of the thousands of microorganisms that live inside of us? What about adult stem cells? "They said we were being too ambitious," remembered James Sherley, who, in addition to being one of the PGP-10, was collaborating with George on the adult stem cell project. "This is the weakness of NIH and I think even NIH itself recognizes it. It's been designed to support work that has a high probability of success. That makes it hard to support things that are truly visionary."[32]

Without exception, everyone I spoke to thought the site visit went extremely well. The dress rehearsal had gone off without a hitch. But that was in front of a small crowd in an advisory role.

Now the real critics would see the real performance with $24 million at stake.

After the site visit, the key was to incorporate NHGRI's feedback into the final proposal before the May deadline, about seven weeks hence. A few months after that, George and a couple of colleagues would travel to Bethesda to answer questions on NIH's turf: a reverse site visit. "A reverse site visit is the weirdest, most unfulfilling experience," said Chad Nusbaum. "You go down there, you put on a show, and then you wait. You don't know how you did. You're on the plane ride home and you don't know if you've won the World Series or not."[33]

George certainly had no idea what the outcome would be, but as he left Maryland, he thought the proceedings felt rushed. An important reviewer had not been in the room and two co-investigators from the PGP team were absent as well.[34]

After discussing an NIH application, reviewers were given the task of assigning it a number from 100 to 500, where 100 was best.* Priority scores of 100 to 150 were most likely to be funded; 150 to 200 might be funded; and proposals with scores greater than 200 had almost zero chance.[35] In mid-December 2008, in the middle of the worst economic crisis since the Great Depression, George sent me an email: "to add to our 1929-style holiday cheer, we just got a 265 score on our renewal."[36]

---

* The NIH scoring system has since changed to a different scale.

Team Church had lost the World Series, and its defeat apparently had nothing to do with open consent. More likely it had to do with open source.

"It does not seem likely that the scale of effort described in this proposal would keep pace with ongoing development efforts by the commercial suppliers," said one review. Elsewhere it called the Polonator "homemade."[37]

George was exasperated and crestfallen. After all the work to make the proposal palatable to NIH and after the triumph of "Gettysburg" nine months earlier, to be called an amateur by his peers was, in the immortal words of Sting, "a humiliating kick in the crotch."[38] George speculated that one of the reviewers, a venture capitalist and former commercial sequencing heavyweight, saw the Polonator as a potential competitor to private-sector sequencing companies just beginning to gain a foothold in the marketplace, to say nothing of those that were still in the development stage. He wondered if the reviewers knew that he was on the scientific advisory board of nearly every single next-generation sequencing company in existence; but how could they not?[39] "The point of the Polonator is not to compete with them," he wrote by way of rebuttal, "but to provide an open and common ground to help as many of them as possible. What we are doing is innovation that companies and academics typically don't do (for example, open source). How do we know if open, platform-independent genomics will continue to lead in cost-effectiveness if we don't fund the experiment? It is clear that it doesn't take a monopoly to create a market environment of fixed high prices and 'proprietary' (that is, secret) components marked up by 80x (Agilent) or 1000x (Illumina) without the community knowing it. It would be rather convenient for the companies to eliminate the main voice for open-source sequencing (and set an example for other would-be open-source grantees)."[40]

He decided to call in one of the big guns, someone the genomics community was not inclined to ignore. This person wrote a stern missive to NHGRI on George's behalf, pointing out that not only was George's team responsible for the only open-source sequencer in the world, but that it had been the driving intellectual force behind two of the current commercial platforms. The email called on NHGRI to grant George's request for interim funding;[41] George wound up submitting a request for some of Obama's stimulus money to tide the PGP over. And he submitted another renewal proposal in 2009. In both cases NHGRI helped guide George through the process.[42]

I recalled a conversation with George in the summer of 2008, not long after Navigenics entered the market and things began to heat up; George was "invited" to Washington by the Secretary's Advisory Committee on Genetics, Health, and Society (see chapter 4) to defend the personal genomics company he founded, Knome. Representatives of deCODEme, Navigenics, DNA Direct, and 23andMe were there as well.[43] The meeting was civil but pointed. New York and California had already come down on the companies, and some committee members were clearly underwhelmed by what the companies were offering to a naïve public and the idea that they had the temerity to charge money for it. In George's eyes, this skepticism was reminiscent of his dealings with NIH.

"When the government protects people from lethal drugs, they look like good guys. But when they protect people from information, they're dangerously close to the people who felt that Gutenberg's was a bad idea because the printing press gave people information that they shouldn't have. That was information about the world. This is information about *themselves*."[44]

Nearly two years later, New York gave a grudging yes to Navigenics, though it demanded that the company operate as a clinical laboratory[45]—I suspected it still might be easier to get medical marijuana in Schenectady than a genome scan. Meanwhile, some genetic epidemiologists were still insisting that information could be toxic and that, like Tom Cruise in *A Few Good Men*, we couldn't handle the truth.

But I had already met plenty of folks who weren't waiting for anyone's permission to look at their own DNA. And I would meet more.

9

"You Can Do This in Your Kitchen"

I followed Kay Aull and her bicycle into her first-floor apartment in Cambridge, Massachusetts. The vibe was pure college student: broken doorbell, peeling paint, dubious plumbing, and a thick smell of cat. At the end of the long narrow hallway was the kitchen. Dishes filled the sink; a few picked-over carryout containers sat on the counter. The linoleum was a bit sticky. After putting her bike and, in a show of modesty, her dirty laundry away, Aull filled me a glass of tap water. She was tall and wore glasses. She seemed both to delight in and be slightly self-conscious about her own transparent nerdiness. If there were such an archetype as "an MIT girl" (she graduated with the school's first biological engineering degree in 2008), Aull might have been one.

My visit to "Chez Kay" was not really about slovenly grad students, their languid felines, or wayward pizza boxes. It was more about the molecular biology laboratory Aull kept in her closet.

Around the time she graduated, Aull learned about a contest sponsored by io9.com, the irreverent science fiction website whose manifesto proudly declared:

> Earth is full of people who want to sell you cheap ways of seeing the future. They tell you tomorrow will be more of the same, with shinier toys. Or that work as we know it is about to end. io9 is the visionary watchdog who calls those charlatans on their shit. We're going to show you a new world that's shockingly different from what you're used to. And it's not always going to be a shiny happy place.[1]

The contest called for entrants to build a real synthetic life-form using MIT's registry of standard biological parts[2] known as "biobricks," which were essentially bits and pieces of DNA known to perform specific functions: turning a gene on or off, binding a protein, etc.[3] What the io9 judges wanted were folks who would avail themselves of these "parts" and make something functional with them. In other words, they wanted their contestants to be biohackers.[4] And what were those? "People who build things in their closet that could eventually do things like replace kidneys or eat oil spills," Aull told me.[5] She built a biological binary counter: just as digital electronic devices use ones and zeros to transmit information or signal events, so too could cells. Aull designed and constructed exactly such a cellular gizmo using biobricks along with bits and pieces she cobbled together herself.[6]

Aull didn't win the free trip to Hong Kong to attend the 2008 synthetic biology meeting. But as first runner-up, she managed to impress the judges with what a young scientist could do with five hundred dollars, a couple of months' worth of evenings and weekends, and a small bit of closet space.[7]

DNA sequencing had long been falling in price by orders of magnitude; so too had countless other pieces of molecular biology paraphernalia. This enabled Kay to emulate many of the things "real" labs did every day, such as perform the polymerase chain reaction. PCR was arguably the most important molecular biology laboratory technique to emerge in the last twenty-five years.[8] As we saw in chapter 5, it allowed one to take vanishingly small amounts of DNA (such as what one might find at a crime scene) and copy it a billion times. This yielded micrograms of DNA, which may not sound like a lot, but is a quantity that was much, much easier to manipulate and visualize. PCR essentially mimicked the way our cells replicated DNA, only on a larger scale. It was absolutely critical for synthetic—and most other—laboratory biology.

To review, PCR works like this: The template DNA sample (that is, the one you're interested in) is heated to near boiling. This separates the two strands of the DNA molecule (remember Watson and Crick's double helix—heat makes it unravel). Primers—short pieces of artificially synthesized single-stranded DNA (about twenty bases long)—are included in the reaction and attach to the separated strands of the template; primers serve as pointers that show the enzyme that does the copying where to get started. The temperature is lowered and that enzyme (DNA polymerase) comes along and begins making copies en masse. Within two or three hours, the process is done. A PCR machine is basically nothing more than a heating block: it gets very hot to separate the strands (about 95–98° C), cools to allow the primers to find their mates (about 45–65° C), and then a little warmer to let the enzyme do its Xerox-like business (about 70° C).[9]

Theoretically, then, if one had some DNA building blocks, primers, and enzyme, one could do PCR with a hot plate, a bucket of ice, and a thermometer. But Kay Aull was unwilling to be *that* much of a do-it-yourselfer ("I don't want to stand in front of a stove for three hours").[10] Instead she found a 1990s-era PCR machine on eBay for fifty-nine bucks. When we met she was very excited because her new gel box had just arrived.* She had ordered this block of Lucite with electrode inputs on the side from DIYBio San Francisco, whose umbrella group aimed to "help make biology a worthwhile pursuit for citizen scientists, amateur biologists, and

---

* By exposing DNA fragments to an electrical charge, it is possible to separate them by size in an agarose gel. Small fragments run faster than larger ones.

DIY biological engineers who value openness and safety."[11] The orange gel box, about the size of a thick airport paperback, looked not much different from what I had used in graduate school in the 1990s. "What's so special about this one?" I asked, turning it over in my hands. Aull's eyes widened. "These are an order of magnitude cheaper!" Spoken like a true student of George Church, which of course she had been.[12]

The rest of her setup was not nearly so high-tech. For an incubator she used a Styrofoam box and a heating element. For a microcentrifuge she stuck a handheld drill through an empty yogurt tub: squeeze the drill trigger and the tub would spin. For enzymes she availed herself of free samples. When the synthetic bio start-up she worked for changed locations (Codon Devices, cofounded by George Church and since reorganized[13]), Aull pulled a tube of enzyme and "a couple other things" from the trash.[14]

With the synthetic biology contest over, she was keen to move beyond the esoteric aspects of DIY biology and launch a project with more intuitive appeal, one that might attract both experienced biohackers and newbies. She chose human genetic testing. "I wanted a protocol that would allow other DIYbio users to move away from prepackaged kits and to try an actual project," Aull said. "In my family people think of these tests as magic. They're not—you can do this in your kitchen."[15]

Aull had a personal reason for exploring the subject, too. Her father suffered from hemochromatosis, a disorder that causes the body to absorb more iron than it can store and can take a severe toll on the heart, liver, and pancreas. Hereditary hemochromatosis is an autosomal recessive disease: carrying two copies of a mutation in the HFE gene usually (but not always) causes the body to absorb too much iron. One in two hundred Caucasians carry two mutations; as many as one in eight carry a single mutation. Treatment is easy, safe, and effective: it's essentially bloodletting—a pint every two to four months for life. The key, however, was to catch patients before their organ damage got too severe. In most cases of classic hereditary hemochromatosis, symptoms didn't begin until the thirties or forties, although there are rare juvenile and neonatal forms of the disease.[16]

If Aull had two copies of her father's mutation, she would most likely need phlebotomy treatments at some point. She told me she would also feel the need to inform her mother, to whom she had not spoken since she was a toddler (Aull was twenty-three when we visited). Would she even know where to find her? "I know her name," she said without emotion. "That's what Google is for."[17]

I asked her if her roommates had any objections to her domestic molecular explorations. "They said I can't keep bacterial cultures in the fridge," she reported. "Which is reasonable." The household reached an accord whereby Aull could keep DNA samples in the freezer, which she did, collected in tiny propylene tubes in a small Tupperware container.[18]

Her roommates might turn out to have been more forgiving than Big Brother. In 2009 Aull and several other members of DIYbio received phone calls from Monitor 360, a consulting service whose clients included governments, nongovernmental organizations, and corporations.[19] Monitor 360 had been asked by a three-letter government agency ("You could probably guess which one," said Aull) to prepare a report on biohacking.[20] In a post on the DIYbio Google group, Aull noted that her interlocutors were particularly interested in how she dealt with regulatory issues. "I told them the truth: I don't have permits, and I don't think the [local regulatory] agencies would have a clue what to do with me if I applied for them."[21] As it happens, Cambridge was the first city to regulate the manipulation of DNA.

The one modest concession to regulation—and self-preservation— Aull made was to set up a limited liability company.[22] The biological supplies she was using were ordered through and delivered to her company rather than to her. In post-9/11 America, this was probably a good idea. One morning in 2004, University of Buffalo art professor and "bioartist" Steve Kurtz called 911 because his wife would not wake up. Hope Kurtz died in her sleep of heart failure. Had it ended there, the story would have been merely tragic, but not all that uncommon. Instead it took a turn for the farcical, the surreal, and the Kafkaesque. When paramedics arrived at Kurtz's home, they found the lab equipment and bacterial cultures he used to make art with the Critical Art Ensemble, which explores the intersections among art, critical theory, technology, and radical politics.[23] Kurtz and Bob Ferrell, one of my former genetics professors at the University of Pittsburgh and the one who obtained the harmless bacterial strains for Kurtz, were both indicted for wire fraud and mail fraud. Kurtz told *Nature*, "The FBI was very thorough about going round to all the cultural institutions and labs we worked with and intimidating and threatening them. It had the effect of almost classifying science to make sure it's further alienated and pushed away from the public." Kurtz was finally cleared four years later.[24] Nevertheless, Kay Aull considered "l'affaire Kurtz" to be a cautionary tale about the perils of DIY. Rather than face the feds, "I'd rather spend seventy-five bucks to become an LLC."[25]

The occasional query from law enforcement notwithstanding, DIYbio continued unabated. The PGP's Jason Bobe cofounded it and remained an active participant: he helped to start the BioWeatherMap initiative along with George.[26] The BioWeatherMap was essentially a community-based environmental genomics effort that was just getting its act together in 2010. The idea was this: Ordinary citizens in different cities and towns would go out and swab public surfaces—doorknobs, hand railings, elevator buttons, dollar bills—with Q-tips and send them to a laboratory. The lab would then extract microbial DNA from the samples and sequence each one. The sequences would be used to identify which organisms are present and presumably what their public health implications are. Much like, say, a pollen count or a pollution index, any given BioWeatherMap will change over geography and time.[27]

During our visit Aull came across as more of a sober realist than a mad scientist. In its embryonic state, the limitations of molecular biology as a hobby were still profound. She noted that the local DIYbio meetings were marked by chronic arguments about safety and lab space. For the former there was no standard; of the latter there was simply never enough, at least not for underemployed amateurs. If she had any more disposable income, Aull told me, she would try to move beyond the closet laboratory paradigm. "I don't have much physical space here. I need a place where it's out of my roommates' way and my cat can't eat it." She explained that Jake, her cat, had tried to eat one of her gels. "He didn't get very far—just enough so that it was full of cat hair. Cats are curious like that."[28]

In the coming months Kay's homegrown results indicated she did not have hemochromatosis but was a carrier. "That was the outcome I was hoping for," she told me via email.[29] She confirmed it by getting tested through DNA Direct, a commercial genetic testing service.[30] The results meant that she felt no obligation to contact her estranged mother. They also suggested that her father was indeed her biological father. "This could have ended with many more awkward conversations than it did," she wrote.[31] "I was prepared to have them, but I'm glad I don't have to."

It was early December, the cusp of cold and flu season in Northern California. Hugh Rienhoff, a geneticist turned consultant, turned to his young, slight, Spider-Man-loving pixie of a daughter, Beatrice, and said, "Beazle, I really think you ought to get a flu shot."

"No," she said calmly. "I don't like shots. But . . . if you want some of my blood for DNA, that's no problem."[32]

A few days later, Beatrice turned five, though at twenty-eight pounds she could pass for much younger.

I had met father and daughter seven months earlier at Penn Station in Baltimore, an enormous and beautiful old Beaux-Arts edifice defiantly standing over a part of Charm City that had seen much better days. I stood just outside the massive revolving doors and watched as a good-looking fiftysomething man in a dress shirt and wire-rimmed glasses carried a blond child in a bright yellow raincoat in one arm, umbrella in the other, through a cold May downpour.

Hugh had brought Bea here to see Hal Dietz, her doctor at Johns Hopkins. Hugh knew the landscape well: he came from a family of Baltimore doctors. His father, Hugh Sr., broke the line by becoming a metallurgist and wanted his son to resist medicine as well; Hugh Jr. could not. In the early 1980s he was a genetics fellow at Hopkins under the tutelage of the father of twentieth-century medical genetics, Victor McKusick.[33]

When Bea was born, Hugh had long since traded in his full-time clinical and lab-bench vocations for biotech entrepreneurship and business consulting. "Working as a consultant is another way of saying you're unemployed," he told me. His current venture was called FerroKin Biosciences, a start-up developing a treatment for iron overload in anemia patients who had undergone multiple transfusions. He had kept his medical license current and for a while did pro bono work at an HIV clinic in San Francisco; now he volunteered at the city's Department of Public Health. But since 2004 his passion—his obsession—had been trying to figure out what was wrong with his daughter.[34]

When Rienhoff's wife, Lisa Hane, was pregnant with Bea, their third child, she was forty-two and at higher risk for having a baby with a chromosomal abnormality like Down syndrome. Her chromosomes—pictures of chromosomes are a fairly crude view of one's DNA—looked normal. Lisa's doctor looked elsewhere: one of the common ultrasound tests obstetricians run near the end of the first trimester is a nuchal scan, which measures the amount of fluid behind the neck of the fetus. More fluid is associated with a higher likelihood of a chromosomal problem and/or heart defects.[35] Bea's scan showed a high level of fluid. At nineteen weeks the couple got an echocardiogram to look for major heart problems; the doctors did not observe anything unwarranted on the ECG.

But with Bea's emergence from the womb came the first moment of recognition for her father. "I saw her feet," Hugh said. "Marfan syndrome flashed through my mind."[36] Marfan syndrome is a connective tissue disorder that affects multiple organs; patients may have enlarged aortas, severe nearsightedness, cataracts, and a swelling of the sac around the spinal cord, among many other features, including the long feet that Hugh noticed on Bea.[37] A person with Marfan is typically tall and thin; there has been speculation that Abraham Lincoln might have had it.[38] Without surgical aortic replacement, Marfan patients' aortas may eventually weaken and rupture, leading to sudden death.[39]

Bea had other physical peculiarities, too, though they didn't look like Marfan. She was floppy, she had a port wine stain (a large red or purple vascular birthmark) on her face, and her fingers were contracted. "I knew there was something going on," Hugh said. "Were these things isolated or part of a syndrome? Was it genetic?" He tried not to think about it.[40]

Lisa was mostly oblivious until the three-month visit to the pediatrician, who told the couple that something was up. Bea was still floppy and she was not gaining weight. She would nurse but would consume only tiny amounts at one time. There was discussion about inserting a feeding tube. Lisa's lowest moment came when a friend visited. "I said, 'Do you think she's okay?' My friend said, 'She's just . . . *really small.*' The way she said it made me think it was bad. When an ordinary person gave her true assessment of what she saw . . . Bea just looked so weak."[41]

She did not appear to have classic Marfan syndrome: she lacked certain features, such as ocular problems. And she had others not typically associated with Marfan, such as severe muscle weakness, widely spaced eyes, and failure to thrive. The Rienhoffs went from specialist to specialist on the West Coast. Each one poked and prodded at Bea, ordered tests, and offered tentative diagnoses, none of which were satisfying to Hugh. One intriguing suggestion was that Bea had a form of Beals-Hecht syndrome, a rare Marfan-like condition characterized by crumpled ears, long fingers, contractures, and scoliosis.[42] Hugh contacted Rodney Beals, who described the syndrome in 1971. But based on Bea's hyperextensible limbs, he didn't think she had it.[43]

Hugh reasoned like a clinician: his daughter needed a diagnosis. A diagnosis would suggest a management plan. A management plan would, he hoped, lead to weight gain for Bea and alert him and Lisa of what to be on the lookout for. Simply getting a muscle biopsy or ordering another round of tests was not a management plan. He was frustrated.

"I thought she had something in the Marfan family. She had long fingers, long feet, and a caved-in chest. That was a place to start. But I wanted her to have a good old-fashioned physical exam. She had to be seen by someone of the old school."[44]

He brought Bea to medical geneticist Dave Valle, someone he knew slightly from his Johns Hopkins days. Unlike the other docs they'd seen so far, Valle gave Bea what Hugh called "a great exam. He knows his syndromes. He knows that if the patient has inverted nipples, then he'd better check the fat on the bottom of the butt. Things like that. Dr. McKusick called clinical geneticists 'the last generalists in medicine,' and I think that's true. That's Dave Valle."[45]

Valle brought in two colleagues, Bart Loeys and Hal Dietz. Loeys was a genetics fellow; Dietz had been at Hopkins since his pediatrics residency twenty years earlier. He was part of a group that had identified fibrillin-1 as the causative gene in Marfan syndrome in the 1990s and probably knew more than anyone on the planet about the genetic basis of the disease.[46] Like Valle, Loeys and Dietz examined Bea for clues to her condition. They looked at her uvula, the small piece of flesh that hangs from the palate. Hers was split: a bifid uvula, something that occurs in anywhere from 1.5 to 10 percent of newborns.[47] They examined her widely spaced eyes and listened to her heart. When they were done they told Hugh that Bea needed an echocardiogram as soon as possible. As it happened, Hugh had already ordered one, thinking that perhaps a defect in blood flow between the chambers of the heart was causing Bea's failure to thrive.[48] But Loeys and Dietz wanted an echo for a different reason. They were worried about Bea's aorta, the largest blood vessel in the body. If she had a Marfanoid aorta that was enlarged and weakened, then it could tear and kill her.

Okay . . . but if it were agreed that Bea *didn't* have Marfan syndrome, then why the urgency? Loeys and Dietz gave Hugh a paper they had just published on manifestations of a Marfan-like disease described by and named after them: Loeys-Dietz syndrome. Loeys-Dietz patients had mutations in a gene in the same biochemical pathway as the mutations that caused Marfan syndrome, the transforming growth factor beta (TGF beta) pathway.[49]

I asked Hugh how he felt at that moment, knowing that a possible diagnosis for Bea had been found. "Terrible," he said. "I was depressed. I read the paper and saw that the average age of death was twenty-six or twenty-seven. Loeys-Dietz was *much* worse than Marfan. I wasn't expecting that Bea might have catastrophic vascular disease. Sometimes you want a diagnosis . . . but really you just want your daughter to be okay."[50]

Hugh brought Bea home and nervously watched the cardiologist perform the echo. It was completely normal . . . The family could exhale. Meanwhile, the Johns Hopkins team sequenced Bea's copies of the two TGF beta receptor genes that had been associated with Loeys-Dietz syndrome. They were clean. Hugh pondered these data: a normal echo, no mutations in the Loeys-Dietz genes, and the presence of extreme muscle weakness, a feature not seen in Marfan or Loeys-Dietz. In his mind, all of this added up to three strikes against these two syndromes. The best news was that Bea did not have any signs of life-threatening vascular problems. The bad news was that Hugh was back to square one.

He went into hypothesis-generation mode. He read everything he could about the TGF beta pathway. Perhaps Bea's phenotype was also the result of something gone awry in TGF beta signaling. But whatever it was, it would have to account for her muscle weakness. Bea could certainly walk, but anyone could see that climbing stairs was a challenge—she would use her hands to help propel herself (to descend, she often opted to slide down the stairs on her backside). Bea's underdeveloped musculature led Hugh to a biochemical pathway related to TGF beta and centered around a protein known as myostatin. Myostatin's normal function is to limit the growth of muscle.[51] Mice that have been engineered without a myostatin gene develop extremely large muscles: they are known as "Schwarzenegger mice."[52] A few champion athletes have been found to carry myostatin mutations.[53] Hugh reasoned that if Bea's muscles were *overproducing* myostatin or something like it, then that might explain her weakness. Moreover, myostatin shared at least one receptor with TGF beta. "The more I looked at it," he said, "the more I thought this pathway was a credible way to get what Bea has."[54]

He began asking around for help. He talked at length to one of the people who cloned the myostatin gene in the 1990s.[55] This scientist thought Hugh's TGF beta hypothesis was reasonable, but he was reluctant to sequence Bea without Institutional Review Board approval—Bea was a human, after all, and entitled to the protections offered all human research subjects, even if Hugh was her dad. This meant that research results from a lab that was not clinically certified could not be returned to research participants or their families. Other requests from other laboratories were met with similar polite refusals.

"I thought, well, shit, I'll have to do it myself," said Hugh.[56]

He found surplus stores in the Bay Area that sold used lab equipment. He bought a PCR machine. He ordered primers and began setting up PCR experiments to amplify the relevant parts of the activin receptor genes that were known to interact with myostatin (activin receptors receive signals that tell cells to grow, divide, differentiate, and/or die). He was just about to buy a cheap DNA sequencer when a friend stopped him and said, "You're crazy. Just send the samples to a core lab at a university. They'll do it for four dollars per reaction." He did.[57]

Within a few weeks sequence data began coming back. Hugh transferred it all to Word files on his computer. He then went to every genomic database to find all the DNA variants that were known in the activin receptors. Did Bea have any that had not been reported? If so, did Hugh and/or Lisa carry them? And if they did, then why were he and Lisa healthy?

After scouring the three genes he'd had sequenced, Hugh found a variant in Bea that had not been reported in other published human genomes (of which, admittedly, there were few). But he had still not had himself or Lisa sequenced. It seemed like an obvious experiment—so why not do it? And why not sequence the myostatin gene itself?

"Because by that time I'd kind of lost interest," Hugh said. "I had a management plan. Bea would get regular echocardiograms. And for treatment we had losartan."[58] Losartan is a generic, FDA-approved drug that is commonly used to treat high blood pressure. For the last few years, it has been used to treat Marfan patients[59] thanks to a fateful Google search performed by Hal Dietz.

When I suggested this version of events to him, Dietz laughed. "That's not the whole story." He came across as a mild-mannered and affable guy; he wore a rumpled sweater and small rimless glasses and had a receding hairline. When I said that I thought he had devoted his professional life to

Marfan syndrome, he gently corrected me again. "Certainly Marfan is an important part of what I do, but it's not the majority. I would categorize [my work as being dedicated to] anyone who has a problem in the first three centimeters of his or her aorta."[60]

In 2001, Dietz's lab found that mice with Marfan mutations developed emphysema, something that occurred in about 10 percent of Marfan patients. The conventional wisdom was that the emphysema was the result of deterioration of the patients' lung tissue over time. But Dietz's team noticed lung problems very early in the Marfan mice's development, which meant there must be another explanation. They suspected that too much TGF beta was the culprit. Dietz's hypothesis was that fibrillin-1's normal function is to keep TGF beta in check in the extracellular matrix, the scaffolding that supports our cells and is a defining feature of our connective tissue. In Marfan patients, the thinking went, fibrillin-1 is disabled and TGF beta runs wild, leading to enlarged aortas and other problems in places where fibrillin normally keeps TGF beta tamped down, such as in the lens of the eye. The Johns Hopkins group showed that antibodies to TGF beta could prevent heart valve problems and aortic aneurysms in Marfan mice.[61]

But could they also prevent lung disease? This was an obvious follow-up experiment, but Dietz wanted a more clinically relevant way to suppress TGF beta. Like, say, with a drug. One of his postdocs did some digging and learned that several blood pressure medications also inhibited the growth factor. Dietz then went to Google and typed in "TGF beta antagonist, FDA-approved." Losartan was at the top of the list. At six months, Marfan mice given a placebo or a traditional beta-blocker had severe aortic aneurysms; they were in bad shape. The losartan-treated mice, on the other hand, looked completely healthy. The drug had actually *reversed* the aortic damage and had positive effects on lung tissue and skeletal muscle.[62] "It was beyond anything I could have anticipated or hoped," Dietz told *Science*.[63]

In some cases, discovering that a drug has profound effects in mice might be sufficient to get a grant funded or drive a stock price up, but rarely does it have immediate ramifications for patients. Losartan was different: it was already FDA-approved and had been used safely and effectively in millions of people, including children, for nearly two decades. On top of that, it was used to lower blood pressure, which was already known to be a good thing in Marfan patients. Dietz and his colleagues began offering losartan to a small number of Marfan kids with severe aortic problems who had not responded to other drugs. A preliminary study by the Dietz group on eighteen kids with Marfan looked promising: once on losartan, their rate of aortic enlargement slowed dramatically. A full-blown clinical trial began, but it would be years before definitive results were in; meanwhile the buzz about the drug was so strong in the Marfan community that some patients opted not to participate for fear that they would be assigned to the control group and not receive losartan. Some obtained losartan prescriptions from their own doctors.[64]

For the Rienhoffs, even though Bea did not have Marfan, losartan offered hope. "Two things were compelling to me," Hugh said. "First, we had not ruled out that Bea was at risk for vascular disease. She will be getting echocardiograms for a long time to come. Losartan actually *arrested* the vascular disease of the Marfan mouse! Second, I looked for cases where people taking this family of drugs had an increase in skeletal muscle mass; I found a bunch of papers supporting that. It seemed to me that, with this drug, Bea could get prophylaxis [prevention of aortic problems] on one hand, and therapy [an increase in muscle growth] on the other."[65]

Dietz, who tended to err on the side of caution, was somewhat lukewarm to the losartan-for-Bea idea, but he was prepared to go along with it given Hugh's research and rationale. But it wouldn't have mattered: Hugh had already made up his mind. "I didn't ask for Hal's permission. I only asked him for the correct dose."[66]

I drove to Hugh and Lisa's place about thirty miles south of the San Francisco airport in San Carlos, a bedroom community with Spanish street names and two-story houses on windy culs-de-sac. Their old farmhouse seemed more Virginia than California—hardwood floors, high ceilings, and a wraparound porch. They had three kids: Colston, then age ten; MacCallum ("Mac"; age seven); and Beatrice. When I showed up in December right before Bea's fifth birthday, no one was home; a neighbor let me in with little to no suspicion. Inside, amid the artwork, children's furniture, and tasteful decor, stuff was strewn everywhere: books, toys, mail. I felt right at home.

An hour later Hugh showed up wearing a bow tie—his doctor outfit, I reckoned, though he only saw patients a couple of days a month. He greeted me warmly, kicked off his shoes, and insisted on carrying my bag upstairs. We retired to his carpeted office in the attic: the Bat Cave, he called it. Piles of paper, notebooks, journal articles, and books littered the floor. Pro-Obama and anti-Bush stickers adorned the wall; Lisa had been a socialist labor organizer when she and Hugh met in the 1990s. Another sticker said, "Evolution is God's Intelligent Design." The room featured two computers. One was for all of the mundane business of being a serial entrepreneur. The other was for "My Daughter's DNA," as Hugh called it, or what I will call, for the sake of brevity, "Project Bea."[67]

Management plan or not, Hugh had not been able to let go of Project Bea. Since I'd last seen him he had resumed his molecular detective work in a massive way. Sequencing three genes was nothing. Now, loaded onto his computer, he had tens of thousands of gene sequences from himself, Lisa, and Bea. Each had given blood. From those samples, someone at Illumina, the market leader in the new DNA sequencing technology, had isolated white blood cells and from those extracted RNA and enzymatically converted it to its DNA form known as complementary DNA, or cDNA.* The resulting collection of genes from Bea and her parents were

---

* RNA, as you'll recall, is the intermediate between DNA and protein; it is the "messenger" that instructs the protein synthesis machinery exactly what to make. Genomic DNA—the 3 billion base pairs you inherit from each parent—is full of extra stuff that doesn't code for protein. If you are interested in which genes a cell or tissue is actually using, you would want to look at RNA, nearly all of which does something useful. First, however, you would enzymatically convert it to its DNA

their three transcriptomes: the sets of all of the *active* genes (transcripts) produced in a particular type of cell, in this case white blood cells. Transcriptomes were easier to work with than whole genomes because 1) they represent only 1 to 2 percent of the genome, "only" 60 million plus base pairs instead of 3 billion; and 2) they include much if not most of the "business end" of the genome: the parts that instruct the cell what proteins to make.

So now Hugh had a set of sequence data from all of the genes that were expressed or "switched on" in Bea's white blood cells.* In this case, that meant about fifteen thousand distinct bits of RNA, about two-thirds of which coded for protein. And he knew to *what extent* any given gene was expressed in those cells. "Why would that matter?" I asked. Hugh pointed to the glowing computer screen where he had aligned his, Lisa's, and Bea's data—a seemingly infinite series of colored peaks and valleys measuring gene expression. In most cases, the peaks were of similar height in all three. Although the conventional wisdom at the time was that humans' gene expression profiles would vary widely, Hugh's family suggested otherwise—the diagonal lines showing levels of gene expression were almost identical for the three of them, even though he and Lisa were not related—certainly not closely. But the patterns were not perfectly correlated, either: Hugh pointed to a variant in a gene he was scrutinizing that had two versions (alleles), G and C. He and Lisa both had one of each: they were both genotype G/C (heterozygous). Bea, on the other hand, inherited a C from each parent: she was genotype C/C (homozygous). So what, I thought. "Ah," Hugh said, anticipating my skepticism. "Look at the difference in expression between the two alleles." The peak

---

counterpart, complementary DNA (cDNA). Why not just sequence RNA directly? It can be done, but people generally don't because it is very fragile. The half-life of RNA is usually measured in minutes. And it's very easy to contaminate. DNA is much hardier and easier to work with.

★   Admittedly, if one or more genes responsible for Bea's condition were *not* turned on in white blood cells, one would never see them. This is clearly a limitation of transcriptomics.

representing the G allele towered over the tiny peak representing the C allele; the G version was expressed tenfold more strongly than the C version. Bea, therefore, appeared to make a tiny fraction of this particular protein compared to the amount her parents made.[68]

Hugh would have to chase down this lead as he would thousands of others, one gene at a time: checking for differences in expression levels, looking for new mutations that neither he nor Lisa carried (he had found several possibilities), looking for known pathogenic variants, trying to guess whether variants he saw that had not yet been reported might be capable of causing Bea's condition, and all the while enlisting pro bono help from academic scientists who analyzed gene expression data for a living. And there was a *mountain* of data: instead of setting his computer's sequence analysis filters to be extremely stringent, Hugh kept his filters loose and inclusive, lest he miss anything among the millions of base pairs scrolling by. It seemed like an impossible long shot.

He smiled. "That's why I call it hand-to-hand combat."[69]

As we sat in the Bat Cave, Hugh picked up a child's acoustic guitar and began to play a sure-handed version of "Fly Me to the Moon." Later we harmonized on "I Will" by the Beatles. As the kids ate dinner in the kitchen, Hugh sat on the floor munching carrot sticks and telling me about his decision to become a conscientious objector during the Vietnam War. I reckoned I liked him because he came across as a smarter, better-looking version of what I imagined myself to be on my best days: a dedicated father, an enthusiastic dilettante, an iconoclast, and a bit of a troublemaker.

A couple of years earlier, at the National Marfan Foundation Annual Conference in Palo Alto,[70] not far from his home, Hugh went to a workshop intended for people and families who had Marfan-like features but lacked a diagnosis. He saw it as an opportunity to represent Bea's point of view and to learn something. There were a fair number of patients and a doctor, Dianna Milewicz of the University of Texas Health Science Center in Houston, who served as interlocutor.[71] Each patient told his/

her story, many of which involved frustration with the medical system. Many went something like this: "I have long fingers, a heart murmur, and the insurance company won't pay for the echo." Or: "I have a high arched palate and narrow feet, long fingers, my father died at age thirty-five, and my physician will not order the genetic test for Marfan." Hugh found it heartbreaking to encounter so many people stuck in the same netherworld of not knowing what their symptoms meant, and without his financial and scientific resources.[72]

Finally it was his turn. "I have a daughter," he began, "and she has some of the important features of Loeys-Dietz, but she doesn't have any evidence of vascular disease. I reasoned that maybe she had a TGF beta problem, or that she had a problem in some other member of that gene family. So I sequenced a few of her genes."[73] At that point, according to Hugh, Mile-wicz stopped him, turned to the rest of the participants, and said that none of them should ever attempt this—it was just too difficult.[74]

Hugh was taken aback. "Who the fuck is this person?" he thought, shocked at the paternalism of a fellow doctor, especially given the current, pre-reform state of health care. "I've never seen a better example of a physician being contemptuous of a patient's own initiative. I live in a world where the patient-doctor relationship is a *collaboration*. That's how I was trained. These people need a diagnosis. Without a diagnosis they can't get coverage. And sometimes once they do get a diagnosis they're screwed anyway because suddenly they have a preexisting condition. But getting a diagnosis is the foundation of everything else: prognosis, management, reimbursement . . . and understanding."[75]

It was an epiphany. People had no idea what genetics and genetic testing were about and the medical establishment didn't seem all that interested in telling them. Hugh was. He went home and wrote down everything he'd done for Project Bea. And he began talking. "If you can make a good soufflé, you can sequence DNA," he told the *Economist*.[76] He and Bea graced the cover of *Nature*.[77] They appeared in *Make,* the do-it-yourselfer magazine.[78] In 2009 they were in *Wired*.[79] A presentation Hugh gave showed up on YouTube.[80]

And he launched MyDaughtersDNA.org, a website that pointed visitors to various accounts of Bea's journey and invited parents, patients, and physicians "to describe for the other users of the site a case that needs some help." A few of them have told their stories and shared their expertise.[81] A geneticist in St. Louis read a description that a Bulgarian man had posted of his twelve-year-old daughter who lacked the ability to cry and suggested that she might have an extremely rare endocrine disorder known as Triple-A syndrome. She did; shortly thereafter she began hormone therapy. "To see a complete stranger solve another person's problem," said Hugh. "That was nice."[82]

But the broader goal was to inform. Everything he'd written down went on the website: how to do PCR, how to pick primers, where to get something sequenced, how to navigate genomic databases. "I felt that just understanding what the steps were would demystify the process. Maybe no one would actually go from beginning to end, but people could understand every little step along the way. A lot of people don't have the benefit of seeing a geneticist, let alone know what it is a geneticist does."[83]

But just as Milewicz considered the Rienhoff approach a bad idea, so too did Bea's own doctor, Hal Dietz. "I could imagine some parents diverting financial resources away from physical therapists and other health-care professionals in order to access the great promise of genomics and sequencing. I could imagine tragic consequences to that decision. I think Hugh is uniquely prepared to understand and deal with the complexities of his daughter's situation. But I can't imagine parents with no scientific background really knowing what to do with the information they would get."[84]

Hugh deferred to Dietz. "When the discouraging word came from Hal, I immediately switched from being the 'bad boy of genetics' to being Bea's dad who did not want to jeopardize her relationship with her doctor, which at that time was tender and new. I'm not condemning Hal. I just decided that maybe I wouldn't tell the world how to make an A-bomb and blow up academia. And besides, it's all out there already anyway, right? I was just consolidating the information. And putting a human face on it."[85]

Eventually Dietz came around . . . at least somewhat. When Hugh and Bea came to Baltimore in 2009, Dietz fully endorsed keeping her on losartan. And when Hugh brought him reams of transcriptome data, Dietz did not roll his eyes; Hugh hoped that this would be the beginning of yet another collaboration on Project Bea. He speculated that his "Tom Sawyer" approach to science might finally be working.[86]

When Hugh called from his cell phone the following summer, he was still pursuing any number of hot candidate genes. But he was even happier that Bea was thriving, if not putting on weight. Her echocardiograms continued to come back clean. She was being a champ about remembering her losartan pill and taking it on her own.

And she was living the life of any other happy five-year-old. She had started to read and learn math. She attended science camp with her older brother Mac and came home ebullient about the experiments they'd done; one sensed the pride in her hypothesis-generating old man. Maybe she would make it to Hopkins as a student one day instead of a patient.

Meanwhile, Bea and Mac had become close: Mac treated her like a peer, teased her as only a big brother can, and stuck up for her when necessary. "Bea's not above kicking a boy in the *cojones*," reported Hugh. "She got into a fight the other day at camp with a boy who was much stronger than she is. Macky had to step in and defend her. She's got to learn to take what she dishes out. But . . . chivalry is good, too, I suppose."[87]

Hugh had gotten off Highway 101, the perpetually clogged artery in and out of Silicon Valley, and we began saying our good-byes. Where was he headed?

"My ambition is to eat lunch. After all," he said, "you gotta have ambition."[88]

10

"Take a Chance, Win a Bunny"

The first PGP-10 meeting was both baby shower and brain-storming session. George expressed his gratitude to us. The other seven of us who could attend sat around a table and made speeches, wish lists, and plans for our genomes. Filmmaker Marilyn Ness was there to capture it, having scraped together enough money to shoot documentary footage of us for a few hours. After a day of offering paeans to George and getting to know one another, fanciful musings, and gentle arguments, five of us adjourned to George and Ting's unassuming Brookline house with the pond in the back for food and drink. The next day we scattered, full of ambition and hope for the project.

The second annual gathering was intended to be both a bigger deal and more substantive: PGP-10ers would see their data fresh from the Polonator, they would consult with a clinical geneticist about it, and they would make decisions about how much of it they were willing to share going forward. The press would be invited to learn about the PGP, its participants, and how genetic information need not be toxic.

It was just after 5 A.M. on Sunday. I sat in the Raleigh-Durham airport anticipating a nap as I waited to board an American Eagle flight to Boston. It was hard to believe that the last two years of running around the country had brought me to this, the presumptive crucible of my entire personal genomics experience. Amy Harmon, the Pulitzer Prize–winning *New York Times* reporter who covers "The DNA Age" for the paper, had called me nearly every day for the last week to ask me questions for a story that would coincide with the Big Event: Why did you do this? What do you hope to learn? What does your family think? What would you redact? I danced around, but I didn't really have any hard-and-fast answers for her. To my own surprise, I was torn. Should I let it all hang out, as I had said I would over and over with studied insouciance, and thereby demonstrate my solidarity with the other nine and my loyalty to the open-consent ethos? Or should I heed my colleague Bob Cook-Deegan and all the other cautious pre-Facebook, pre-Twitter people I knew and opt for the old privacy norms?

At the beginning of my involvement with the PGP, I had longed for my genome to be something more than someone else's lab experiment, more than a set of anonymous data, more than an *abstraction*. The lesson I failed to heed, of course, is to be careful what you wish for. But before revisiting that theme, a little concrete backstory.

In 1976, at age forty-two, after months of denying the existence of the growing mass in her chest, my mother was diagnosed with breast cancer. I was twelve and don't remember it well, perhaps because I was such a burgeoning pothead at that age. I do recall fidgeting in the backseat of my parents' Plymouth Valiant and listening to my older brother, whom I idolized, trying to impress upon me the gravity of the situation: "Mom could *die*." For once I knew he had to be wrong.

My mother opted for a radical mastectomy. Twelve years later, after her doctor found a thickening of cells in her remaining breast, she went back to the hospital for another one. Years later, when I asked her about

this decision, she smiled and said, "I already had one breast gone, so I figured why not make it even. If you don't have breasts, you can't get breast
cancer."[1] When it comes to her own health, she has always been a practical woman (less so when it comes to her children's and grandchildren's
health).

Given her young age at diagnosis and her heritage, there's a decent
chance my mother carries a mutation in either the BRCA1 or BRCA2
gene. Five to 10 percent of breast cancers in white women are due to aberrant versions of one of these genes. In Ashkenazi Jewish women with hereditary breast cancer, three mutations in particular account for as much as
90 percent of the disease burden, perhaps more. Females with one of these
mutations have an 80 percent lifetime risk of developing breast cancer.[2]
The chances of me or my brothers developing breast cancer were fairly
remote: about 1 percent of U.S. breast cancers occur in males. We are
"hormonally hostile" to breast tumors.[3] But these genes are not on the sex
chromosomes; thus, men can *transmit* BRCA mutations to their offspring
just as well as women. I could have passed it to either or both of my two
daughters, who turned seven and ten in 2009.

At the dinner banquet that closed the first annual Cold Spring Harbor
Personal Genomes Meeting, I sat next to the doyenne of breast cancer
genetics herself, Mary-Claire King. She was and is a brilliant thinker and
tough-minded professor at the University of Washington. She gives off
maternal charm and empathy. She is a tireless crusader for human rights;
in the 1980s, her lab used molecular genetics to prove kinship among
Argentine families in which the parents had been imprisoned by the
military dictatorship and the "orphan" children were being adopted out
surreptitiously.[4] In 1994, King lost the race for the two major breast cancer genes with such grace and humility that her already-stellar aura only
grew.[5] Years later she would find dozens of novel mutations in those two
genes, mutations that Myriad Genetics, the company with a monopoly
on genetic testing for breast cancer, had missed.[6] I told Mary-Claire my
story and asked for an on-the-record interview with her, but she quickly
recognized that, with the PGP meeting less than two weeks away, what

I really wanted was genetic counseling. And she was happy to give it. She suggested that I redact several million base pairs around BRCA1 and BRCA2 before going public; if my BRCA carrier status weren't public then I would not have to lie to my daughters. If I didn't have a mutation, then no harm, no foul. If I did, then Ann and I could talk to the girls when they were older. It all sounded so simple.[7]

Ann and I agreed that at the very least, I should find out whether I carried a breast cancer mutation before putting my entire genome on the Web. If one or both girls were at high risk for developing breast cancer later in their lives, we wanted them to hear it from their parents when they were ready, and not from some intrepid blogger while they were still in elementary school. I had already discussed my risk with Elissa Levin at Navigenics, but her company was not offering BRCA mutation results—she could only quote me risks based on my family history. Thus I hoped the next PGP trip to Boston would bring some clarity to the whole business.

But without sequence data neither this nor any of the other promised moments of genomic truth could happen at the October gathering, and back in Polonatorville, all was not well. At the end of 2007, the machine was reportedly ready to ship. By the time of the Marco Island meeting the following February it had become a source of conversation, fascination, and maybe even buzz: an open-source sequencer from George Church's lab that was one-third the cost of the next cheapest commercial competitor! But the truth was, at that time the Polonator was more about promise than it was decoding actual nucleotides from people's genomes. It was a good-looking, well-engineered box that didn't have working software: Mr. Spock without a brain.

Rich Terry, a boyish, thirty-four-year-old engineer from Boston, and grad student Greg Porreca—short, close-cropped hair, New Jersey accent—had devoted much of the previous year working seven days a week trying to breathe life into the Polonator. Terry spent his days tweaking design problems, thinking about microfluidics (extremely tiny channels that carried chemicals in and out of the machine), and, together with

Porreca, writing code that would get the moving parts to go where they needed to be at each step. Had Terry known that this is what his life was going to be like, I wondered, would he still have taken the job toiling in the Church lab? He paused . . . and then he groaned . . . and then he laughed. "Noooooo!" he said. "I don't know . . . Last summer was tough. I like to ski, though, so I don't mind being indoors in the summer."[8]

I asked him and Porreca about other next-gen sequencing technologies. They handicapped the field and more or less stuck to the conventional wisdom of the day: Illumina/Solexa was the tentative front-runner, 454 Life Sciences had squandered its early lead by being too expensive to run, and Helicos charged too much for its instrument and was late to the party. Both also said that ABI's SOLiD platform was essentially an earlier version of the Polonator—the primitive tangle of wires hooked to a microscope, named after Simpsons characters, and known as the D.05—in a box.[9] "It's very similar," agreed George.[10] In one sense, this was true: in the early 2000s, Agencourt Personal Genomics licensed the key Church-lab patent that described the nuts and bolts of polony sequencing, which meant sequencing short stretches of DNA in massively parallel fashion using the enzyme ligase, whose raison d'être is to stitch together DNA fragments.[11] ABI, the dominant purveyor of the Sanger sequencing technology that had conquered the Human Genome Project a few years earlier, was in danger of falling behind up-and-comers 454 and Illumina; it had to make a move. After evaluating more than forty next-generation sequencing technologies, the company in 2006 settled on George's sequencing-by-ligation; it bought Agencourt for $120 million.[12] Kevin McKernan, who was CEO of Agencourt and subsequently became senior director of scientific operations at ABI[13] (which itself merged with Invitrogen in 2008[14]), said it was a logical choice given the promise of the method and its source. "George's mind is always in a lot of different fields at once. He pulls from nanofabrication, microfabrication, and computer technologies. He often has very simple ideas, but they are fundamentally different from the way other people are thinking about things. And his lab got the proof-of-principle done." McKernan conceded the Polonator machine was similar to the SOLiD instrument, but emphasized that the "chemistry is very different."[15]

When I put the same question to Chad Nusbaum, he laughed. "Is SOLiD a D.05 in a box? No! But I can understand why George's people would say that: they have a slightly biased position. There are similarities, but SOLiD's molecular biology is much different. I told George I wanted to call his machine Ligation-Inspired QUery into DNA, or LIQUiD. He said, 'Well, it's open-source. It can have as many names as you want.' He was neither particularly amused nor especially annoyed by my joke."[16]

But George was annoyed about something else. In his mind the Polonator's open-source status remained its most defining feature. And through Agencourt's dealings with his lab, George felt that the company had revealed its true mercenary stripes. "Before they licensed our technology, they said, 'Oh yeah, we like this whole Red Hat–Linux thing: let's go with it.' And almost the instant they got the license from Harvard, they started saying, 'Well, maybe we're not so keen on this open-source stuff.' And after they were purchased by ABI, it was like, 'We're definitely *not* interested in open source.' And they dropped our license just like that. I think they didn't like the 'open' part and they're hoping we'll go away."[17]

The Polonator couldn't go away, however, if it never arrived. The Max Planck Institute in Berlin had one. Jeremy Edwards, a former Church trainee, had one at the University of New Mexico. Another lab at Harvard took a flyer and committed the requisite $150,000+ to buying one.[18] In early 2009, George said that "it's not going fast but it's going okay,"[19] which I took as a fairly sober assessment from its co-inventor and tireless booster. Meanwhile, despite the arrival of newfangled software courtesy of Porreca and Terry, and despite the impassioned insistence of a number of genomics movers and shakers that an open-source sequencer was a good thing to have in NHGRI's portfolio and that the Polonator should be taken seriously (see chapter 8), Nusbaum admitted over lunch in 2009 that his team had still not done much with it. The Broad already had a cavernous building full of Illumina and SOLiD machines plus a Helicos unit fresh from the factory floor down the street. "It's a good machine," Nusbaum said of the Polonator. "But I have no ambitions for it at the moment. It's not obvious to me what I'd use it for right now. . . . I think that would be an awkward conversation to have with George." He looked sheepishly at his soup. "I have to prepare myself that it will eventually happen."[20]

But if the Broad didn't need the Polonator, perhaps the Polonator didn't need the Broad quite yet. In 2010 Kevin McCarthy admitted that his company, Dover, was almost certainly fooling itself when it launched the Polonator two years earlier. There had been problems with the manifold that moved the reagents in and out, and as a result the machine was leaking. Its self-cleaning mechanisms were not up to snuff. And the software was becoming increasingly byzantine. "Eventually the Church lab and I agreed that the fluidics and the software needed to be blank-slated," McCarthy told me. Dover swapped out the manifold in favor of leakproof rotary valves.[21] And while Greg Porreca left Team Church in 2009 for a start-up,[22] in New Mexico Jeremy Edwards stepped in to advance the biochemistry and develop new sequencing protocols. In a few months Dover was shipping Polonators again. McCarthy, who had spent a fair amount of time traveling around and troubleshooting machines in the field, made a user manual available in 2010.[23] By that time, there were fourteen Polonators installed. "That's probably more than Helicos," George pointed out[24]—and for the moment he was right.[25]

Now it would be a question of winning over some of his own people. "I think my lab is divided," he admitted. "One group is pushing the Polonator into extremes that are ridiculous—they are making it do all kinds of amazing new tricks. For example, you can synthesize DNA on it now; you can use it to look at cell morphology and behavior. But the other half of the lab is skeptical and they aren't using it." He chuckled into the phone. "I suggested we cut off all of the lab's supply money for everything except the Polonator."[26]

Across town in Kendall Square, Helicos was still making a precarious crawl across the Valley of Death. In September 2008 the company said that it anticipated fifteen to thirty orders for HeliScopes in the next nine months.[27] In December it reversed course and announced that it would lay off 30 percent of its workforce and retract any and all past predictions of both its future financial results and HeliScope orders.[28] That same week Helicos realigned its senior management: Steve Lombardi was out as CEO, replaced by board member Ron Lowy.[29] Lombardi would remain president and focus all of his efforts on commercializing the HeliScope. "Stepping down as CEO was hard for me," said Lombardi, though he insisted that "it was the right move to make. We needed to go really, really fast. I couldn't do both jobs."[30]

The news got worse. Helicos's stock, which had peaked at sixteen dollars per share in January 2008, bottomed out in December at twenty-two cents. Two thousand and nine began with the company's first customer, Expression Analysis, deciding to return its HeliScope,[31] which had been delivered to the company's Durham, North Carolina, labs with great fanfare the year before. "The thinking was pretty simple," explained CEO Steve McPhail, a friendly and rugged-looking guy with a firm handshake, of the decision to send the instrument back to Boston. "We had two large pharma clients that had opted to use Illumina data. Both of them said to us, 'You *need* to get into this market.' By then we estimated that Illumina already had over fifty percent of the next-gen market. Whenever that type of standardization happens, there are opportunities for outsourcing."[32] Expression Analysis became the first genomics services provider to be fully certified on every platform Illumina offered—genotyping, gene expression, and sequencing.[33]

Not only was the HeliScope getting lapped by the Illumina and ABI sequencing machines in the marketplace, the company was hemorrhaging cash. By mid-2009 it was down to its last $5 million.[34] Helicos brought in an investment banker to advise on "strategic alternatives," which was construed as a euphemism for trying sell oneself to the highest bidder.[35] Would Stan Lapidus's insistence on an independent Helicos be trumped by the company's need for a lifeline?

"It's inescapable that the other start-ups in next-gen sequencing—454,

Solexa, and Agencourt—all got married to Daddy Warbucks," Chad Nusbaum said in 2008 with some prescience. "Helicos didn't . . . or hasn't yet.
It's no small thing to invent a machine and make it work. But neither is it a
small thing to get that machine out in the field to customers, to support it,
to sell it, and to manage a supply chain. Once you start delivering instruments, it's a whole other ballgame. You're supporting users, the software
doesn't run, there's a bug here, a hardware problem there, yadda yadda.
Every single one of these companies has had problems shipping reagents.
Those are the sorts of growing pains you expect. But big companies like
ABI and Illumina, with limitless resources and closets full of intellectual
property, can keep pulling old stuff from the attic for years. I've always
liked the Helicos people. They're earnest and smart. But they're a small
company."[36]

Tim Harris intimated that Helicos's engineer-centric culture cost it
valuable time in getting the HeliScope out the door. "The scientists knew
how to sequence and the engineers didn't," he said. "The engineers started
over and made every component new; every single part was custom made.
That has a large impact on how fast you can go." A few months after publishing a paper in *Science* in which he and his colleagues used the HeliScope
to sequence a virus,[37] Harris left his post as senior director of research for
an academic gig at the Howard Hughes Medical Institute campus in rural
Virginia.[38] "My departure was amicable, but my role was limited," he said.
"The people who ran Helicos had a specialist view and I was a generalist.
I was in a research pigeonhole and I wanted a larger playing field. It was
time for me to go."[39]

Eventually a few rays of light began to poke through the clouds hanging over Helicos. In the coming months it would raise additional money.[40]
The stock price rebounded from its all-time lows.[41] Steve Quake, the guy
who invented the Helicos technology, sequenced his own genome with a
HeliScope in a few weeks, with a small number of people's help, and for
a reasonable-at-the-time cost of $48,000.[42] Helicos scientists showed that
one could use the HeliScope to sequence RNA directly[43] without first con-
verting it into DNA, something that had never been done before and that

none of the other platforms could do. More new applications emerged.[44] It sold four instruments to a science institute in Japan.[45] A Waltham-based diagnostics company bought a HeliScope and promptly launched a test of three hundred genes related to hereditary cardiovascular disease.[46] A few more orders came in.[47] It wasn't a torrential pace exactly, but it was fast enough to buoy spirits in Cambridge, for the moment anyway.

As we finished our Portuguese lunch down the street from his office, Lombardi waxed philosophical. "We're not out of the woods, but the life of a start-up is a hypothesis. You hope that your technology and product strategy are right, but it's always going to be a tough environment. In the 1990s I was part of the investor relations group at Affy saying, 'Why even *worry* about Illumina?' Now what are they saying about Illumina? 'It's over! They've won! They're invincible!' I don't think that's ever true in this business."

Two months after we spoke, Lombardi was out. On a Friday after the market closed, Helicos quietly filed a form 8-K with the Securities and Exchange Commission announcing his resignation. "Mr. Lombardi did not resign as a member of the Board of Directors as a result of any disagreement with the Company on any matter relating to the Company's operations, policies or practices," it said. His successor as CEO and president, Ronald Lowy, personally thanked him for his " . . . efforts and professionalism to help Helicos achieve its goals." Elsewhere in the document, under a section titled "Future Conduct," there was a standard "non-disparagement" clause—neither side would say anything unpleasant about the other.[48]

Steve was not about to violate either the spirit or the letter of the agreement. "Things happen," he told me a few days after his departure became public. If there were any bitterness in his voice, I couldn't detect it. "It was amicable. It was bittersweet. But it was the right thing to do for the company. Don't worry about me, Misha . . . I'll pop up somewhere."[49]

For Helicos, life remained tenuous. In March 2010, investment bank Leerink Swann dropped coverage of the company, citing concerns over the HeliScope's high price, limited market, and looming competition.

"Time is running out," warned the analyst.[50] A few weeks later the company announced that substantial doubt existed over "its ability to continue as a going concern."[51]

When I landed in Boston for the second annual PGP-fest, I didn't have time to dwell on either the plight of sequencing companies or my own existential problems. My nephew Noam was in the hospital. A senior in high school, he was tall and fit, a rower who had somehow developed a rare form of deep-vein thrombosis, Paget-Schroetter syndrome, a condition that causes blood clots in the extremities of otherwise healthy young people.[52] The docs broke up the clot but in the process gave him too much morphine. Noam's breathing slowed to an erratic crawl. After a rough night, he began to come out of his narcotic haze. He received heparin and would start on other anticoagulants soon, including Coumadin. Coumadin, aka warfarin, is the most widely prescribed blood thinner in North America, but different people respond to different doses and the therapeutic range tends to be narrow—not surprising given that its original use was as rat poison.[53] Warfarin has become the advance guard in the brave new realm of pharmacogenomics (see chapter 6). If we can genotype people for the genes that code for certain metabolic enzymes that break down drugs in the liver, we can predict who will respond to warfarin and at what dose; but we're still working out how much it costs compared to old-school methods.[54] I wondered whether Noam had been genotyped but decided against asking about it as I listened to my agitated brother simultaneously talk at me on his cell phone over the ambient noise of Massachusetts General Hospital, minister to his ailing son, and fumble with the television in the hospital room, desperate to tune in the Boston Red Soxes' bid to salvage the American League Championship Series against the Tampa Bay Rays. I said I would check in later.

Pharmacogenomic testing is often hailed as the genomic tool that will be the first to move from "bench to bedside."[55] That may be true, but the latter phrase still makes my skin crawl. I imagine it has been used in thousands of PowerPoint presentations to describe the arduous path from research laboratory to clinical practice. It is a mom-and-apple-pie sort of goal, an agreed-upon desire of every stakeholder in the personalized medicine field. It is an implicit call for randomized clinical trials of high-tech postgenomic measures of human individuality à la response to warfarin. It is a sort of "consensus statement." It is also a cliché, a shorthand that has helped to spawn countless "road-map" and "translational medicine" initiatives and all sorts of other schemes to make personalized medicine the "standard of care."

But how close were we to that reality? With or without genomics, when would we be in a position to circumvent the inefficiencies woven into the system, to say nothing of our own stubborn humanity: the initial complaints of pain, the visit with the confused and green resident, the referral to the ER, the four-hour wait, the shunting from specialist to specialist, the messy procedure, the middle-of-the-night wakeups, the reams of paperwork, the titrations of so many medications, the crushing disappointment and fear wrought by complications, and the uncertainties of reimbursement? And all of this before we failed to comply with our doctors' orders anyway. Nor were doctors themselves immune from recalcitrance. In his outstanding 2007 *New Yorker* article and subsequent book, surgeon and medical writer Atul Gawande described the quixotic crusade of Dr. Peter Pronovost to have doctors follow a simple five-step checklist to avoid central-line infections in intensive care units.[56] In its first fifteen months, the program saved $175 million and 1,500 lives, despite physician resistance to it. In 2009, eight years after it started, the checklist was finally rolled out in twenty-eight states.[57] Clearly this quick, simple, and inexpensive intervention could make a difference. What about genotyping for drug response? I didn't—and don't—know. What I did know was that if this health-care "system" is what we were getting for 16 percent of our gross domestic product in 2005, and 25 percent of our GDP in 2025,[58] or roughly a zillion and a half dollars, it was pretty obvious that we were getting ripped off. President Obama had his work cut out for him—we all did.

Noam was supposed to be released on Monday but developed another clot and had to go back in for more surgery. He would have at least three subsequent procedures over the next week, including the removal of a rib. When he came home he would have to dose himself with warfarin. His doctors would be watching him closely to be sure he stayed within the therapeutic range; otherwise he could bleed to death.

With the help of his family, doctors, physical therapy, blood thinners, and Vicodin, he regained the full use of his arm over the ensuing months. For their part, the Red Sox won Game 6, but fell to Tampa Bay in Game 7 and were eliminated from the playoffs.

I took the train from the bedside (Mass. General) to the bench (George's lab). It was almost noon and Marilyn Ness and her film crew had been running around for most of the morning. With her dog-eared shooting script in hand, she seemed to be the only one who knew what was going on and where everyone was ("George will be here any minute," she assured us).[59] A *New York Times* photographer brandished two digital cameras with enormous lenses and checked the lighting.

Former Clinical Genetics fellow and now-attending doc Joe Thakuria arrived ready to perform a skin biopsy on me in George's office. He wore a threadbare sweater and wire-rimmed glasses; his olive skin came courtesy of his Indian father and half-Filipino mother. In kindergarten Joe was diagnosed as severely autistic. His parents were urged to put him in special education, which they opted not to do. In Catholic secondary school he vaulted to the top of his class and stayed there. He began college at sixteen and med school at twenty. When he was finishing his residency at Penn in 2001, his younger brother died in his sleep from a cardiac arrhythmia. He was twenty-four. Given his brother's general good health, Joe intuited that his death was likely caused by something genetic. "It seemed clear to me that in the future medicine would (or at least should) be able to screen, identify, and save patients like him." He has since become convinced that his brother had Brugada syndrome, a genetic condition that causes arrhythmias and puts one at high risk for sudden cardiac death.[60] In 2009 he approached his family and proposed exhuming his brother's body to get a DNA sample for testing, but "for emotional reasons" his family had been slow to act on this idea.[61]

Joe lightly numbed my arm with a shot of lidocaine and made a three-millimeter hole near my tricep. A skin biopsy is better than a sharp stick in the eye, but based on my experience, I'm not sure how much better—my arm was tender for almost two weeks. I don't think the procedure was Joe's forte, but why would it be? Clinical geneticists, after all, don't really do "procedures"; they spend most of their time talking to people, just as psychiatrists do. Thus, Medicare and insurance companies were not all that inclined to pay for their services. And this, among other reasons, is why geneticists have long dwelled at or near the bottom of the physician food chain, which I happen to think is a shame.

Joe was relaxed and seemed to radiate the placidity that good physicians do when they're in their element. I thought he would have been beaten down by having had to sort through thousands of genes and decide which ones were important. This meant setting up stringent filters in the software to remove all the noise and the "normal" sequence that did not vary from the reference sequence. I learned later that he had been up until all hours combing through the data and preparing for our consultations—maybe his tranquility was just exhaustion.

Soon George appeared in a brown corduroy jacket, his hair still wet. He turned to me and spoke in a low voice. "You get to watch me fail in real time on a national stage."[62] I smiled at his self-deprecation, though I didn't quite understand what he meant. Ting arrived to sit in on their consult with Joe. People were calm and talked in hushed tones, but a nervous energy was present in the room. Marilyn studied her schedule and gave terse, sotto voce directions to her crew. The "First Family of Genomics," minus daughter Marie, took their places at the small circular table in the center of George's office.

"You have four variants that are most concerning," Joe said to George and Ting, " . . . which are not *really* all that concerning. Of the four, there is a variant associated with susceptibility to multiple sclerosis." Ugh, here we go with the MS again, I thought. "I've enclosed three research papers and one commentary on the specific variant." Ting furrowed her brow. "I would not lose sleep over it," Joe said. They talked about the small sample size of the study indicting this variant and its relatively high frequency (84 percent) in the population. If it were truly bad news and powerful enough to do damage in youngish people, then most of us wouldn't have it—this is both the beauty and the cold indifference of natural selection. George asked if he was heterozygous, that is, had just one copy rather than two. He was. Joe noted that this variant was associated most often with MS in younger females (most MS patients develop the disease prior to age forty; George was fifty-two). And, Joe said, its contribution to MS appeared to be relatively weak. "Is there any MS in your family?" asked Ting. "I don't think so," said George.[63]

Not an auspicious start. They moved on. George—and several of us, it turned out—carried a variant that appeared to cause an increased susceptibility to tuberculosis in West African populations. As with the MS allele Joe had identified in George, the major risk allele in the tuberculosis susceptibility gene had a frequency of greater than 80 percent. But with the possible exception of James Sherley, none of us had obvious West African roots. What was going on here? Joe, it seemed to me, had made his choices about what to discuss with us based on the strength and validity of the science in each case more so than on the likelihood we would actually develop these conditions. Fair enough, but none of these polymorphisms was likely to have a dramatic—or even undramatic—effect on our lives.

Variant number three was thought to be associated with ovarian hyperstimulation syndrome and longer menstrual cycles. Hoo boy. This was starting to feel like an early Woody Allen movie. "It could be relevant to our daughter, though," said Ting, always ready to see the beaker as half full.[64]

She and George bantered about whether there might be mosaicism, a phenomenon whereby some cells in the body are genetically different from other cells. George mused on it, thinking aloud about the prospect of comparing skin to blood. "That's a level of sophistication we're not quite at yet," he said.[65]

George struck me as willing to play his part in celebrating the fact that we were having actual genomic consultations, even though this approach would not be scalable to one hundred thousand. But he also knew that this whole exercise was mostly for show. "The database fields for frequency, penetrance,* and research quality are pretty much empty," he said.[66]

My own consult followed George's and proved no more scintillating than his. George's was such a nonevent that I suppose I should have been suspicious about mine. All the SNPs Joe went over with George and Ting were from the Affymetrix chip harboring five hundred thousand markers. This was information I'd already seen via SNPedia and, to some extent, Navigenics, six months earlier. Joe handed me my folder, asked me if there was anything I didn't want to know, and confirmed that my phenotypic data were indeed my phenotypic data. Yes yes yes, I thought. I'm over-weight and depressed—get on with it! On a piece of paper before me were three markers. They all had "affx" in their descriptors, which meant they were from the Affymetrix 500 SNP chip. Been there, done that. I felt like the bride-to-be at her wedding shower and I had just been handed a pair of lacy underwear I already owned and frankly wasn't sure I really wanted to wear again.

I followed George outside into the hallway between his office and lab. "Your sequencing run failed," he said apologetically. Now I understood the remark about getting to watch him fail in public. "Only six of the ten will get their sequence data today."[67] And, I would come to find out, even that sequence data was of fairly low quality and woefully incomplete. The moment of truth turned out to be a false alarm or, at best, a dress rehearsal. He could see that I was crestfallen (even though I had no right to be) and started to apologize again (even though he had no reason to). He assured

---

* Penetrance refers to how likely individuals carrying a particular version of a gene (allele) are to actually express a trait associated with that gene.

me that all ten of us were in the queue for a full genome sequence from Complete Genomics. "No no, it's okay, George," I said. Marilyn interviewed me and to her I defended George and said yes, I'm disappointed but hey, shit happens. I said I felt "our" disappointment was because the PGP-10's expectations had been raised, because we were so intimately involved with the project, and because I had talked to so many people about the PGP, about the technology, about the ELSI aspects, etc. And because in twenty-four hours the world was supposed to hear all about the glory of seeing our own genomes.

I would have to wait a few more weeks or, more likely, months or years. Knowing that I had SNPs that slightly altered my risk for prostate cancer and tuberculosis was hardly a revelation. Nor was learning that I was a carrier for trimethylaminuria, aka "fish-odor syndrome," although that was interesting in a 23andMe and *Beavis & Butt-Head* kind of way. Because I carried a mutation in the gene encoding the enzyme flavin-containing monooxygenase-3, if Ann were to carry a mutation in the same gene, then our kids would each have been at 25 percent risk of stinking: their urine, sweat, and breath would be replete with aminotrimethylamine, a nitrogenous base that also happens to be a product of decomposing plants and animals.[68] It smells like rotting fish. I admit to being biased (everyone thinks their own kids tend to walk on water), but to my nostrils, my children smelled no worse than any of their peers and almost always better than me. Curiously, though, I can't stand eating fish and am repulsed by the way they smell. More evidence of self-hatred, perhaps. In any case, I was relieved that I'd chosen a freshly scented mate who wasn't a fish-odor carrier: my family dodged a bullet there. Persons with trimethylaminuria really do suffer—some so much so that they commit suicide.

Much later I asked Joe to reflect on what exactly went on during those initial consults. "That was definitely a dress rehearsal in some respects," he wrote. "That date in 2008 had been set before we had any data from your exomes. If I were doing a consult today, I wouldn't even mention any of those high-frequency, probably benign variants. We've gotten a lot better at eliminating that type of noise."[69]

The day after my consult the ten of us met on the third floor of HMS's New Research Building, fifteen months after we'd last met. We sat around the same table: Rosalynn, Keith, John, George, Kirk, Esther, James, and me, now joined by Stan Lapidus and Steven Pinker.

There were smiles all around. I suspect we were all happy to be there, with the possible exception of George. His MO was lowering expectations. But how could he do that with the *New York Times, Boston Globe,* and *Wired* in the room, to say nothing of a sound crew and a cameraman, and a press conference to follow?

He tried. "This is really about listening," he said. "We know there are still a lot of rough edges."

He put up a slide titled "Major Points":

1. Thank you.
2. Today is a start, not a final product.
3. The PGP is research, not a genetics service.
4. We are providing some interpretations, but mainly to initiate study and discussion. Decisions about releasing one's data should be largely based on other considerations.

He put up our mug shots, the ones where we sported pieces of measuring tape on our foreheads. As always, our photos were labeled with our Coriell Institute accession numbers in case anyone wanted to order our cells or DNA.

We went around the table and gave little speeches again. I don't remember what I said other than that I was thrilled. And for the most part I was.

Kirk, still smarting from the sperm donor fiasco, I imagined, compared the arrival of the PGP to the emergence of widespread vaccinations in the nineteenth century and said we should expect backlashes.

Esther: "My feeling is today shouldn't be exciting. We don't understand this yet. We know ninety words of Russian and we've just been handed *War and Peace*." (That's Война и мир to you, comrade.) I agreed with her, but the truth was the PGP-10 had hardly been given the whole of *War and Peace* . . . more like a few dog-eared pages; an outline, or the CliffsNotes at best.

James said he hoped we would try to lead first with education and that people would come to recognize the importance of genomic information—we were all products of it, after all. And with immortalized cell lines, genomes of individuals could now be propagated through time.

Steve Pinker emphasized that his interest was largely scientific: "Behavioral genetics and twin studies have underscored the importance of genes in our psychological makeup. What's missing are links between an individual's genome viewed holistically and that individual's makeup. The answers have implications for brain function and how we evolved. Even identical twins are distinguishable. Why? What is the role of chance? What is the role of ontogeny?"

We talked a little about "crowd sourcing." To what extent would all of these questions be answerable when we had data on one hundred thousand people and the world was free to peruse it?

Robert Green gave a talk about the REVEAL study: Risk Evaluation and Education for Alzheimer's Disease. Despite the fact that many of us had heard his shtick before, his data were so compelling and his delivery so engaging, we all sat there rapt as he told us, once again, that hearing one was at high risk for a devastating, untreatable, late-onset disease was very unlikely to cause permanent emotional or psychiatric harm.

He also said that we were now witnessing the early stages of a war between scientific and unscientific approaches to the genome. "We've already lost a similar war in nutraceuticals," he said. I stole a glance at Rosalynn to see if the founder of a nutrigenomics company would flinch. She wouldn't.

And Green said that yes, genomic information should have the traditional attributes—analytical validity, clinical validity, and clinical utility—in order to be used in medical testing.[70] In other words, genetic tests should measure what they say they measure, they should be predictive of something, and whatever that something is should be treatable somehow. But he also put forth the idea of "personal utility," a phrase he credited to the

University of Washington geneticist Wylie Burke. The personal utility[71] argument goes something like this: "Maybe I don't care whether or not my samples have been typed in a CLIA-certified lab. Maybe I just want to know where my family came from and/or the genetic basis of how my pee smells after I eat asparagus. Maybe I can't do anything about Alzheimer's and maybe the test isn't perfect, but I'd like to know as much about my risk as I possibly can. You may deride those sorts of things as silly or bad ideas, but for me, they represent information I am interested in and they are no one else's business."

Stan Lapidus, the founder of Helicos and the same guy who'd imagined opportunities for "positive eugenics" when I spoke to him a few months earlier, was wearing a bow tie and looked as though he might have stepped out of an earlier era. He asked George if we could review "redaction opportunities."

"We consider these difficult exercises," George said with characteristic diplomacy. "APOE will be in the next set of data you receive. Watson found his out, even though he said he didn't want to. If you redact one thing, it is likely there will be something even scarier later on."

George then reminded us that even if we did redact certain portions of our genomes, anyone from a research institution could order our cells, grow them in culture, extract DNA from them, and then sequence whatever he or she wanted. We would be fully identifiable.[72]

I had heard this refrain both from George and from others when I mentioned the possibility of redacting my breast cancer genes: "Today it's breast cancer," they said. "Who's to say tomorrow we won't find something worse in your genome?" A fair point, but I had a strong family history of breast cancer and a high likelihood of carrying a single gene variant that would strongly predispose my kids to it. Wasn't family history still "the gold standard"? At the press conference that afternoon, Pinker said he did not want to know his APOE status. He and I were the lone potential redactors of anything in our genomes. Everyone else but us gave the PGP carte blanche to put the entirety of his or her data on the Web. I said only that I wanted to see mine before I released it.

But redaction, as it turned out, had practical, catch-22ish consequences. By asking to see my data so that I could decide whether to redact a small piece of it, I had further delayed the day when I would actually see my data. Steve Pinker and I were now at the back of the sequencing line. Jason said it was too hard to segregate our data and our cells. Redaction was not scalable. "Redaction poses logistical issues," he wrote to me. "It's not clear that we can honor such requests and manage all of the upfront costs and downstream consequences right now. We're at a point in our development of infrastructure, protocols, and quality control that, like in the early days of the Model T, you can get it in any color you want . . . as long as it's black."[73]

"Getting sequenced requires being comfortable with the cell lines being distributed and with the unknown part of the sequence being released," George told me. "Watson surprised me by redacting APOE, but everything since then has been unsurprising. I feel that if you're gonna get your genome sequence and make it public and you know a modicum of genetics, then you know that one gene is just the tip of the iceberg. Frankly the PGP-10 were not recruited to get a homogeneous set of opinions. I think if anything we failed to get enough skeptics among you nine. It's a trade-off—if someone is ambivalent, then we're not satisfying the IRB. But if everyone is drinking the Kool-Aid, then we're not getting any useful feedback."[74]

Graduate student Sasha Zaranek, a soft-spoken, Russian-born, and Canadian-raised guy who managed the sequencing data pipeline, said that because of our reluctance to release our data without strings, my sequence information and Pinker's had been encrypted in order to protect it and us; therefore the other eight PGP-10ers would get their data first. "It's just practical for us to do it that way," he said. "The one thing we don't want to do is to get your wishes wrong."[75]

He told me that the PGP had an ongoing internal debate about whether to release rough data often or to release polished data less frequently. The former approach hewed more closely to the PGP ethos of openness. On the other hand, shitty data could actually do more harm than good. Could he give me an example? "Halamka has been blogging about things we think are sequencing errors," he said quietly.[76]

"Listen, I'm not trying to be a dick," I said. "But I was consented eighteen months ago, George long before that, and there's still almost no real, high-quality exome data on any of us." Other groups were starting to publish exomes and whole genomes they'd knocked out in a much shorter time.

"We're *all* frustrated with the pace," Sasha assured me. "We've been working at this for *years*. But as George says, we're eventually going to sequence six billion people, so what does it really matter where we are on this exponential curve?"[77]

Touché. What was my hurry? Was being among the first really so important (other than to my editor, who wanted a draft of this godforsaken book)? I had been naïve in the extreme. When I strolled into Harvard in 2006 and saw the XY and XX chromosomes denoting the men's and women's restrooms, respectively, and began reading about the PGP, going to meetings about personal genomics, and listening to people discuss it in breathless terms, I envisioned the PGP to be the Six Million Dollar Man of human genomics: better, stronger, faster. And maybe it would be someday. Maybe in a couple of years, instead of just the ten of us eating carryout vegan Chinese in a Harvard Medical School conference room, tens of thousands of us would converge on Fenway Park and celebrate the PGP database busting at the seams.

After the meeting most of us went to a bar in Boston to be part of a science café—we each spoke for a couple of minutes, discussed the PGP with patrons, and imbibed (some of us anyway). I asked George what had gone wrong. As usual he was sanguine. "We have this mountain of bad data with people who have no family history for any genetic diseases. So . . . garbage in, garbage out. That was my expectation, but we had to do *something* . . . so we did what we could. Hopefully this will improve before our next installment. It's not clear how many more events like today we have in us without making some major breakthroughs first. I think we have to cure a disease before we get the *Washington Post*, the *New York Times,* and the *Boston Globe* here again."[78]

On the flight home I wondered if maybe this long, drawn-out process of returning our data was an opportunity to rethink the whole thing: Did I *really* want my genome to be available to everyone via a mouse click when I clearly didn't feel that way about other aspects of my life? I was still not on Facebook or Twitter (though I would eventually succumb to the latter) and felt no burning desire to join the social networking crowd—fielding dozens of requests from other 23andMe customers wanting to "share" results and exult in our common Ashkenazic pedigree was more than enough.

My therapist speculated that maybe my whole foray into the land of public genomes was just another form of acting out, a cry for attention, a way to ingratiate myself with "the cool kids" like Esther Dyson and Steven Pinker and George Church. Maybe she was right. My problem was that, at the end of the day, I was *not* one of the cool kids. I was not as smart or unflappable or selfless or capable of seeing the big picture. I was constitutionally ill-equipped to roll with the punches; George, on the other hand, excelled at it. He told me that even as an infant his mother said he was an easy baby—"a stoic," she called him.[79] At the risk of invoking genetic determinism, as a baby I was a howler. So were my kids when they were infants. And as an adult, I was not about to put my Social Security number or the GPS coordinates of my house on the Web. Maybe this meant the PGP was not for me. Maybe negotiating the details of a public genome was more than I, to say nothing of my family, could handle. I resolved to not make any decisions until I learned my BRCA status for the most common mutations in Ashkenazi Jews. But with the Polonator still in the midst of a long, slow gestation and the Harvard sequencing facility having its own problems, the PGP was unlikely to provide that any time soon.

The only way around this was to look elsewhere: to the world of commercial genetic testing, which, presumably, would deliver garbage neither in nor out. I called my local hereditary cancer clinic and explained the situation as best I could (glossing over the fact that I wanted to do this because I was about to, um, put my entire genome on the Internet). I asked if they would order a test for me for the Ashkenazi breast cancer mutations and send the bill ($535) to my insurance company. The genetic counselor on the other end of the phone was pleasant but officious: she told me the clinic staff would be happy to order the test for me as soon as my mother contacted them, answered the same set of questions I had, and then got tested herself. It was my mother who was the cancer patient after all, the counselor explained: if *she* didn't carry a mutation, then the rest of the family was off the hook. It was simply the most efficient way of doing it. I understood this—it was the traditional, Bayesian* way of identifying mutation carriers in families.

But it wasn't what I was after. I explained that my mother was seventy-five years old, had had both breasts removed, and as far as I knew was long done with her ovaries. Why did *she* have to be considered the proband—that is, the one who first sought consultation for a genetic disorder? *I* was the one who wanted the information, and if my insurance company wouldn't pay, then I would. I loved my mother, but there was a whole bunch of reasons I didn't want to make her an active part of my little journey of genomic self-discovery. The counselor politely but firmly refused to budge. "If your mother absolutely won't do it, then we can talk," was as far as she would go. I said thank you and hung up the phone. "Fuck you," I thought. This was exactly the kind of situation that led to the rise of the personal genomics companies.

I thought about pulling "a Hugh Rienhoff": extracting my own DNA, ordering the primers to amplify the BRCA genes, and borrowing

----

* Bayesian probability is a statistical approach that uses some prior probability of an event and then updates that based on whatever new information comes to light. Obviously, if my mother were *known* to carry a BRCA1 mutation, then the odds of me carrying one would jump from less than 20 percent (given her ethnicity and history of breast cancer) to fifty-fifty.

some lab time, space, and equipment. I could have an answer in a week, assuming I could still recognize the business end of a pipette. That was a big assumption—I hadn't done real lab work in years and I didn't want to compromise my scientist friends by skirting the human subjects research rules (the same ones I was supposed to enforce) and infringing the Myriad patents on the breast cancer genes, as odious as I found them to be (more on those in a minute).

Instead I called a friend who worked for DNA Direct, a San Francisco–based company that was an early proponent of providing Web- and phone-based access to genetic testing for people who wanted to avoid wading into the morass of paperwork or otherwise involving their health-care providers. Or their mothers. DNA Direct would submit my test for coverage to my insurer; my out-of-pocket cost would be two hundred dollars for the counseling session (again, I noticed that insurance companies didn't like to pay for medical care that only involved talking).

For me it was money well spent. I didn't have to bug my mom or my colleagues. And my genetic counseling session with DNA Direct was like a warm bath. I sat in my office and talked on the phone with Lisa Kessler, an experienced counselor who had just had a baby and worked from her New Jersey home. Listening to her reminded me of why I was such a hopeless genetic counselor and never even bothered to take the board exam. I was nervous and awkward with patients; Lisa's voice and telephone manner, on the other hand, were quiet and soothing. She was comfortable in her own skin. Her empathy felt genuine. We discussed the PGP, genetic counseling training programs, her newborn, and all of the various breast cancer risk models. She explained what we could and couldn't say about my risk with and without a genetic test for the three Ashkenazi mutations in BRCA1 and BRCA2. Given my family history (which was incomplete) and ethnic background, my mother's risk for carrying a BRCA mutation was somewhere between 12 percent and 38 percent. My risk for carrying one was therefore half of that. Myriad told her that the three mutations I was getting tested for accounted for 90–95 percent of the mutations in the two genes that were found in Ashkenazim.[80]

A few weeks later she called with my results. "I have good news," she said.[81]

"I know," I said. Eight weeks before her call, I had churned two milliliters of my saliva into a plastic tube and sent it to 23andMe's lab for processing. The company had typed 600,000 markers on my DNA and I could now log on and see what those markers had to say about various traits of mine. To my surprise, those traits now included the three major Ashkenazi breast cancer mutations! Why was I surprised? First, my impression from an early conversation with company cofounder Linda Avey was that 23andMe was not interested in typing customers for SNPs that conferred high risks for single-gene disorders. It seemed to me that in the beginning they were skittish about the prospects of having to deliver bad news over the Web. But when I looked at my report online in 2009, I saw my carrier status for mutations not only for the Ashkenazic breast cancer mutations, but for a handful of other single-gene diseases, too, cystic fibrosis and sickle-cell anemia among them (I carry neither). In fact, for months George had been agitating for direct-to-consumer genomics companies to start testing for rare, single-gene disorders. Hugh Rienhoff also had said that by causing such serious phenotypes, nature was telling us that these genes were the important ones and the ones the consumer genomics companies should be paying attention to.

The other reason I was surprised that 23andMe had dipped its toe into breast cancer genetics was the fact that both the BRCA1 and BRCA2 genes were patent-protected and controlled by a company called Myriad Genetics.[82] 23andMe was almost certainly infringing upon Myriad's patents.

Whenever I told someone outside the genetics world that as part of my research I was studying the effects of gene patents on access to genetic testing, he or she usually did a double take. "You can patent a *gene*? How is that possible? Isn't that like patenting a tree? Does that mean some company owns part of me?" I was and am sympathetic to those intuitive suspicions. But holders of gene patents made several counterarguments. One was that they have *not* patented a gene found in nature. They have instead patented a gene that's been cloned, amplified, and/or otherwise manipulated by humans: gene patents, their holders said, cover *inventions*, not discoveries—just like all valid patents do . . . or should. And, they argued, the isolation of a gene was a true invention. Furthermore, proponents would say that they have patented the *method* (or methods) of looking for medically important mutations in those genes. Finally, companies with exclusive licenses to particular genes like to say that they need their particular monopolies to expand the market for testing of those genes.[83] In their view, exclusive patent licenses are necessary to spur innovation: without the promise of a nice payoff at the end, investors would not sink money into commercial genetic tests and patients would be forced to rely on the vagaries and slow turn-around times of research-based testing.[84] In general, Bob Cook-Deegan, postdoc Shubha Chandrasekharan, and I (along with many colleagues at Duke and elsewhere) had found that the last argument did not hold water. Rights to the cystic fibrosis gene, for example, were licensed broadly and the test had always been cheap, reliable, and easily available.[85]

Indeed, genetic screening itself seemed headed toward the "close to free" model. In 2009 a company called Counsyl had sprung up whose product was a "Universal Carrier Screen" for more than one hundred genetic diseases, $349 out of pocket or free to customers with insurance. The company included PGPers/famous-Harvard-guys Steven Pinker and Henry Louis Gates, Jr., among its advisers.[86] The Counsyl tag line: "Thinking about starting a family?"[87] Its website described a "campaign" to end preventable genetic disease,[88] presumably by prospective parents who were carriers of mutations in the same genes not having kids or by selecting against affected embryos.

Preconception genetic screening has improved people's lives immeasurably. That said, for Counsyl to call this a "campaign" akin to the campaign to fight AIDS was a bit over-the-top, I reckoned—not everyone thought of hereditary deafness as a disease, for example.[89] And just as being a sickle-cell carrier is protective against malaria, being a carrier of—and perhaps even afflicted with—other genetic diseases or "disabilities" probably confers evolutionary advantages in certain cases.[90] But Counsyl made no bones about its mission:

It is something new, born of the realization that cutting-edge science and market forces can actually increase equality and promote social justice. It is a cause, a campaign to finally end the needless suffering of preventable genetic disease. And most of all, it is you. Call us idealistic, but we believe that everyone loves their children and will do the right thing when it comes to safeguarding their future.[91]

For Myriad, neither Counsyl's launch (although Counsyl didn't test breast cancer genes) nor the 23andMe move into BRCA1/2 could be construed as good omens. Although Myriad's profits from BRCA testing had been robust, its image had not. Because it had a monopoly and refused to sublicense its tests, and because it charged more than three thousand dollars, Myriad had been called "probably the most hated diagnostics company." (Given Myriad's many years of experience with BRCA testing and its high rate of insurance coverage,[92] whether it was the "most hated" was debatable.)

But the company was under assault on two fronts. First, by returning results on BRCA1 and BRCA2, even on just three mutations out of the more than 2,500 known, 23andMe appeared ready to flout Myriad's licenses, although Linda Avey downplayed the intellectual property aspect. "Because this subset of BRCA markers doesn't fully replicate the Myriad test (and it's not positioned to be an alternative for in-depth BRCA testing)," she wrote to me in 2009, "we're not certain how Myriad will react."[93] But if 23andMe could get away with it, how long before other labs started screening for hereditary breast cancer susceptibility genes?

The year 2009 also brought a full-blown legal challenge: in May the American Civil Liberties Union filed suit against Myriad, claiming that the company's monopoly on the BRCA genes: 1) made it impossible for women to access other breast cancer genetic tests; 2) prevented them from getting second opinions about their results; and 3) allowed Myriad to charge exorbitant prices.[94] Myriad responded by filing a motion to have the lawsuit dismissed, saying that the ACLU lacked the standing to bring the suit and that "a mere policy disagreement" did not warrant a declaratory judgment.[95] Judge Robert Sweet said in so many words that the plaintiffs did indeed have the standing to bring the suit and that he would like to hear the case.[96]

At the time I thought the ACLU suit could be important, but I also thought it might be too little too late: even its critics conceded that Myriad did an excellent job of testing BRCA1 and BRCA2,[97] although the company had been criticized for overselling its test to women who didn't need it.[98] I believed this cash cow would dry up of its own accord: most of the relevant BRCA patents were set to expire within the next few years, as were most other gene patents related to single-gene disorders. But Judge Sweet opted not to wait. In March 2010 he handed down a shocking summary judgment that made the court's view perfectly clear:

> In light of DNA's unique qualities as a physical embodiment of information, none of the structural and functional differences cited by Myriad between native BRCA1/2 DNA and the isolated BRCA1/2 DNA claimed in the patents-in-suit render the claimed DNA "markedly different." This conclusion is driven by the overriding importance of DNA's nucleotide sequence to both its natural biological function as well as the utility associated with DNA in its isolated form. The preservation of this defining characteristic of DNA in its native and isolated forms *mandates the conclusion that the challenged composition claims are directed to unpatentable products of nature.*[99] [emphasis added]

In other words, DNA in one's body does the same thing it does outside one's body: *it carries information.* Thus, a gene in a test tube or a sequencing machine is not much different from a gene in a cell in a human body. It cannot be considered a true invention and therefore cannot be patented. Obviously, the defendants were not happy about this ruling and vowed to appeal.

But in George's view, even if the ruling had gone the other way, Myriad's business model was still doomed. "I think that protecting individual SNPs is legally unsustainable," said George of the Myriad patents. "It's *so easy* to get people that data without infringing. We're just going to sequence whole genomes and give them the raw data. If the companies think that that's infringing their patents, then we'll just give people a machine to do it themselves. And if *that's* infringing, then we'll just give them the parts to make the machine. It's reductio ad absurdum. There's not a lot of sympathy for patenting of DNA sequences. It's something that's at the intersection of the open source and human rights movements: 'This is mine and you can't have it.' "[100]

On a personal level, knowing that I was "clean" for the Ashkenazi mutations via two independent sources was a relief. There was still a chance that I carried another BRCA mutation. Ann said she could live with that risk if I could; I thought I could. In the two years since I'd been consented, the world had made small but significant steps in the direction of less genetic privacy: dozens of people were in the process of getting sequenced and even some research participants were receiving their own genetic information. The first six whole genomes had all made their data public. By 2010, the wholesale cost for a full sequence at high coverage was less than $10,000, which meant a lot more people would soon be in the pipeline. And judging by the twelve thousand prospects who had already signed up to join the Personal Genome Project,[101] the destigmatization process appeared to be well under way.

Of course, in the final analysis none of that really mattered: this was *my* genome to do with what I wanted. I had already made my phenotypic information public: if the world cared, it could read my profile and know that I was prone to anxiety and depression. Was it likely that I carried anything in my genome more stigmatizing than that? Of course, any of tens of thousands of researchers could now test that hypothesis by ordering my cells ($85) or my DNA ($55),[102] sequencing me eight ways to Sunday, and publishing any damning allele I might carry. As I've said, this was not something I was terribly concerned about, but until now no one had ever had to give it much thought one way or the other.

When I moved to Durham, I had no idea how to find a doctor. I'm still not sure I do other than by asking my white-coated colleagues. To find a general practitioner, I wound up making a list based on somewhat arbitrary criteria: she had to be reasonably close to my house, an internist, and accept my insurance. I preferred female physicians: I don't know why, perhaps because of some deep-seated mommy issue. Other than a Club Med in Grenada, I didn't much care where she went to school or did her residency or even if she came highly recommended—I assumed she would be competent, what with Durham being the "City of Medicine" and all. I had never really thought of my doctor as "someone to talk to." She was someone who made me stick out my tongue, whacked me on the knee, and told me to turn my head and cough, which I did with great awkwardness. I found one who was affiliated with Duke and she was terrific—pleasant and sharp, and with a Ph.D. in biochemistry even. We chatted about drug development as she looked in my ears. But not long after my second visit she moved to Singapore. Argh. I had to start over. I got my insurer's provider directory and started googling. This process yielded not much more than addresses and phone numbers—it seemed like it led mostly to commercial sites that, for a fee, would tell me my prospective doctor's "grade" or "score." Based on what, exactly? Choosing a physician was not like choosing a restaurant. Or maybe it was.

I finally found one through her practice's website—it was done up in nonthreatening pastels and had pictures of happy-looking women bouncing on medicine balls and little boys flying balsa wood model airplanes. So far so good—it seemed like so few primary care docs had websites . . . maybe they wanted to perpetuate the black-bag-and-tongue-depressor stereotype. I went to see her and we talked. She was young and Australian; a vegan (they're everywhere!) who leaned Republican. When I told her what I was doing with my genome, I expected shock, chastisement, probably befuddlement. But she just said, "Wow. That's really cool."

"Do you have any advice for me?" I asked. She twirled her stethoscope around her finger and looked thoughtful. "I'd stock up on long-term care and disability insurance," she said. I laughed nervously, but she was dead serious. She had had enough experience with insurance companies that she didn't want to make it easier for them to deny me coverage (this was before the passage of the Genetic Information Nondiscrimination Act, for what that was worth).

I appreciated her concern and I saw the logic of her argument: if we knew our own genome sequences and could gauge our own risks, then we could adjust our coverage accordingly. But I wasn't doing this to game the system. In my mind, buying extra policies would be giving support to the fallacious idea that our genes are deterministic and we should be afraid of them. I couldn't square it with the ethos of the PGP, to say nothing of my own convictions. And besides, until now insurance companies hadn't really given a rat's patootie about genetics. What better way to get them to start caring than by hoarding coverage?*

And for all of my neuroses, I found that I really couldn't get too worked up about any of the other Churchian nightmare scenarios coming back to bite me. I had a hard time imagining anyone planting my DNA at a crime scene, shunning me because I was descended from an infamous villain, or, say, my employers deciding they didn't like what they saw on my paternal copy of chromosome 11 and firing me. My genome would

---

* Robert Green told me about speaking to a meeting of long-term care insurers and explaining to them that people who learned they were APOE4-positive (and therefore at higher risk for Alzheimer's) were five times more likely to report that they had purchased long-term care insurance. "The underwriters immediately recognized that this would allow for adverse selection and some grew quite agitated," said Green via email. "They made it clear that if the American public wanted an insurance product like LTC insurance, they would have to have a 'level playing field' in which insurers were not disadvantaged." (Email from Robert Green, January 25, 2010)

be, I presumed, pretty far down any real or hypothetical list of reasons for my termination. As for my kids, even if I didn't trust my wife implicitly (I do), I don't need a DNA test to know that my kids are mine: they have my wife's good looks and common sense, but my nose, my perverse sense of humor, my bad teeth, and at least some of my existential anxiety. That's not genetic determinism, that's just life. And given that I had made it to age forty-five, I didn't believe I would discover anything earth-shattering in my genome about my own health anyway, and if I did, I didn't think that that knowledge would ruin my life—on the contrary, it might even improve it.

Yes, the PGP was a leap of faith. I remembered Baylor's Amy McGuire discussing the uncertainty implicit in agreeing to have one's entire genome sequenced.[103] But as the Estonian philosopher Hermann Alexander Keyserling wrote, "Faith, like courage, rests on consent to uncertainty."[104] When my father and I play Scrabble and I put down a word for a lot of points, but in doing so set him up for a Triple Word Score, he often says, "Take a chance, win a bunny." Sometimes I get away with it; other times he makes me pay with a well-placed Z, X, or Q. The PGP was, in some ways, a microcosm of our lives: we were all consenting to uncertainty and incurring risk. We were all trying to win a bunny. I had come this far and I was still curious. I wanted to see it through, to walk the walk. From here on in, I would redact nothing.

I wrote to the PGP brain trust: "PGP #4 is in."[105]

11

"Something Magical"

Paper Japanese shades were suspended by long strings over the sliding glass doors in the back of George and Ting's house, which meant they covered only the bottom part; early winter morning sun streamed through the top. No overhead lights were on; the light in the room was natural except for the heat lamps under the table against the wall where the family turtles basked (as a child, Marie was an avid turtle breeder and it remains an abiding Church-Wu pastime). Out in the backyard the pond was partially frozen, the mulberry trees were bare, and snow was everywhere. It was barely 7:15 but Ting was already gone to the lab to tend to her mutant fruit flies.

George, who was waiting at the door when I arrived, had been up for hours. He described himself as "in sync with the sun" and "very circadian," although in fact he was more than that: he was narcoleptic.[1] At one point during our first big PGP event, as most of the PGP-10 sat around the table with cameras rolling, he fell asleep. For a moment it was almost imperceptible: he just sat back in his chair, head still upright, eyes closed. The rest of us stole furtive glances at one another . . . should we wake him? Ignore him? In a couple of minutes he was back as though nothing had happened. "The big disadvantage of narcolepsy is the social stigma," he told me. "People get insulted when I fall asleep in the middle of a one-on-one conversation. And it does compromise me somewhat: I always miss some part of lab meeting. People can wake me if they want, but they usually don't. That's just how our society is set up.

"My daughter is just like me. What's really funny is when she and I go to the sleep clinic with my wife. Marie and I will both be asleep halfway through the conversation," he said.[2] "It's the same at night: I tell Ting that I'm so keyed up I don't think I'll be able to sleep—and then boom, before my head hits the pillow I'm out."[3]

"We do want to figure out what's going on with our sleep, especially with Marie, since she's trying to get through school. I already did the flunk-out thing; we want to try to keep that to a minimum in the family."[4]

He filled a coffee cup with water at the sink and handed it to me. His relationship to food was as idiosyncratic and indifferent as his relationship to sleep. I asked him whether he chose not to eat animal products for moral or dietary reasons. "Neither," he said. "For biochemical ones." He ate once a day. He claimed to like the taste of food but, as far as I could tell, never experienced deep pangs of hunger. Yet at 245 pounds he was hardly wasting away. "I can go for days without eating," he shrugged. "Or I can eat to excess."[5] I got used to seeing him at meals where he was a bored spectator while those around him cleaned their plates with gusto.

We sat in the corner of the Church family room/dining room, him at his computer desk and me on the couch behind him. He was working on a manuscript and, in doing so, indulging his other scientific passion: synthetic biology. One of his long-term objectives was to engineer a genome from scratch. Craig Venter's team had already taken a bacterium and systematically removed its genes one by one in order to identify the minimum number needed to sustain life. Venter hoped to synthesize the first manmade bacterium, *Mycoplasma laboratorium*.[6] But George was undertaking something different: he wanted to tinker with the genetic code itself—that is, to rewrite it so that the three-letter DNA "words" would code for different amino acids than they do in nature. One could see the geek appeal of this, especially to an inventor: make something completely new and build new parts with which to do it. That said, one could imagine practical applications for invented genomes. If crops were to use a novel genetic code, for example, then they would be of no use to pathogens: every such crop would immediately become virus-resistant, even to viruses that had yet to be discovered. Jim Watson had said more than once that "someone has to play God,"[7] and indeed, he often volunteered; but George was always a bit more reticent. "It's kind of weird," he said, "this idea of making something impervious to viruses we know nothing about."[8]

As ever, he was driven by advances in sequencing technology, and as ever, his drive seemed to me untainted by any of the Buddhists' Three Poisons: ignorance, hatred, or desire.[9] When I asked how important it was that his lab be the one to win the X Prize, his response was basically "not at all." He was nominally affiliated with almost all of the seven registered competitors in the contest to sequence a hundred genomes in ten days for less than $1 million, though only a victory by his "Personal Genome X-team" and its fleet of Polonators could actually bring him the glory. He insisted it didn't matter. "I am totally unconflicted. I hope everybody wins." The bottom line, he said, would be the bottom line: whoever could sequence genomes on the cheap would get the brass ring. Illumina had become the dominant next-generation sequencing platform, in George's lab and everywhere else. "The most cost-effective one and the one with the longest reads [among the short-read platforms] is Illumina," George said. "We use Illumina for almost everything. It is the one to beat or improve upon."[10]

But he hoped that that improvement would happen soon. When, in 2008, Complete Genomics announced that it would begin sequencing whole human genomes for five thousand dollars in the coming year, George practically clapped his hands with joy. "I'm delighted! This changes the game. Since 2004 sequencing has been getting ten-fold cheaper every year. Can we squeak out one more factor-of-ten reduction in one year? I sort of doubt it . . . but that would be pretty exciting! My principle is this: I don't want to sequence my genome or anyone else's until the price is right. Now I think the price is right. And nobody's suing anybody! Isn't that great?"[11]

For a moment, it was. But Complete Genomics got blindsided by the financial collapse just like everyone else—the retail five-thousand-dollar genome would have to wait until at least 2010. Meanwhile, in 2009 Life Technologies (the entity formed by the merger of ABI and Invitrogen[12]) sued Illumina, which promptly countersued.[13]

As a young scientist in 1987, a few years after Tito's death during the waning days of the Socialist Federal Republic of Yugoslavia, Rade Drmanac (RAH-day Dur-MAHN-itch) submitted what he assumed would be a doomed grant application to the U.S. Department of Energy proposing a novel approach to DNA sequencing. He was a young guy from the wrong side of the Iron Curtain petitioning Reagan's America for research funds. But to his utter shock, DOE loved it—he got the money. "One hundred fifty thousand in real U.S. dollars!" he remembered. "In a communist country and from the same DOE that builds atomic bombs!" Among the swooning reviewers of his proposal was another young scientist, who wrote, "This is the best theory ever proposed for sequencing a human genome." Drmanac told me that even though his then-brilliant idea was premature (the necessary camera and computer technology did not yet exist), he still recalls George Church's lofty praise with pride.[14]

It turned out that getting the grant was the easy part. Drmanac worried that Yugoslav officials would keep the DOE's dollars and pay the award in dinars, whose value was diminishing by the day; the country was on the brink of hyperinflation. There was also concern that the government would skim some off the top. The solution, Drmanac and his fellow scientists reasoned, was to keep the money on U.S. soil. "We opened an account in the U.S. embassy in Belgrade," he said. "So the dollars never went out of there. We would bring in bills and invoices and embassy officials would pay them from our grant. Not a single dollar was lost."[15]

Drmanac has a stout profile, a thicket of gray and black curly hair, a boxer's nose, and a ready smile that accents his crow's-feet. When he's not wearing them, his glasses are hooked into the V-neck of his sweater. As Yugoslavia continued its inexorable disintegration, Drmanac moved to the United States and worked at Argonne National Laboratory in suburban Chicago for a while before starting his first company. He doggedly pursued DNA sequencing technology until the Human Genome Project caught fire and then switched to gene *discovery*, which was all the rage, but something he considered to be a distraction—"the gold rush," he called it. Once the first complete (and expensive) human sequence was in hand and it became clear that finding genome-based drugs was not going to be a cakewalk, the investment climate for DNA sequencing technology became hospitable again: we needed more genomes if we were ever going to understand them. Around this time Drmanac met Cliff Reid, an easygoing software entrepreneur who happened to have degrees from MIT, Harvard, and Stanford. The two crafted Drmanac's technology into a business plan and set out with a simple but tantalizing message for VCs: "We are trying to achieve something other sequencing companies are planning to do in ten years. We are not interested in releasing a product that is only ten percent better than what's already out there. If we cannot develop a method in two years' time to sequence a complete human genome for less than $10,000, then we will have failed." By 2007 Drmanac and Reid's new company, Complete Genomics, had raised more than $12 million—not nearly enough, but enough of a vote of confidence to keep going.[16]

Not only did the company have new technology and new money, it had a novel business model. Unlike 454, Solexa, ABI, Helicos, and a smattering of other would-be players in the suddenly wide-open sequencing space, Complete Genomics did not want to get into the instrument business. "Market penetration for new machines is *so* slow," said Drmanac. "Early on we realized that even if we had an instrument *today*, people would not be able to use it." Or want to, necessarily. In a previous company he'd started, Drmanac had developed a small chip upon which one could sequence several thousand base pairs of DNA—a few genes' worth, which was a lot at the time. His thinking then was that every lab would have one on every bench. "But when we started talking to people, they said, 'No, we prefer to send our samples out to be sequenced.'"[17]

Many if not most life scientists believed that their time was better spent actually *doing science* than learning how to operate a new machine that was likely to become obsolete in a couple of years. Pharmaceutical companies, for example, mired in a long slump and hit hard by the economic downturn, had made it clear that certain expenditures were on the chopping block: half-million-dollar toys would no longer be purchased with impunity.[18] Nor were would-be customers excited about the massive quantities of ancillary data that were generated with every complete sequence.[19] Complete Genomics' DNA sequencing product would therefore be a black box: DNA samples in, sequence out. And the company would store genomic data for customers at its headquarters. It was a pure fee-for-service play: genomes at wholesale.[20]

By 2008 investors and board members were anxious. Complete Genomics had been operating in stealth mode for months, which can feel like a long time in Silicon Valley. Company insiders pressured Reid and Drmanac to demonstrate proof of principle to the world: "Sequence *E. coli*," they urged. The two founders resisted. "Success in bacteria doesn't guarantee success in the human," said Drmanac. "From day one we were going to be sequencing the *human genome*! I was tired of sequencing small organisms. I said, 'Human genome. Let's do it.'"[21]

In July 2008 they did it, and reportedly for less than four thousand dollars in materials.[22] Eschewing peer review, a few months later they revealed many of the details on their own website. Drmanac described it to me in simple terms. "Take human genomic DNA, randomly fragment it, make arrays, hybridize, ligate, and sequence." The arrays were regular microscope slides with a billion tiny slots, each of which could accommodate a single wad of balled-up DNA: a nanoball. Unlike other next-gen technologies, it required no beads or other substrates: just pure DNA. A single slide could yield seventy gigabases of DNA, or the sequence of a single human genome twenty-three times over, and with an error rate of less than 0.1 percent.[23] By September 2009, Complete Genomics had sequenced fourteen whole human genomes.[24]

When I visited the company's headquarters it looked to be typical of Redwood Shores, California: an angular mix of salmon-colored brick and mirrored glass surrounded by lush greenery. It sat just beyond the fringe of the Google campus. Inside was a sort of chaotic feng shui: a trickling fountain drowned out by leaf blowers and lawn mowers outside the windows, cushy furniture, boxes piled up in the hallways. The kitchen was lavishly appointed with free drinks, multiple microwaves, and two refrigerators. Instead of an office, Drmanac had a cube—a big cube, but still a cube. There was a "dot-commie" feel to the place. "I think it's a very healthy culture there," said an admittedly biased George. "It's one of innovation and an amazing amount of openness, especially considering their business model is as closed as it could possibly be."[25] The marketing director was wary of me, which was probably appropriate. She gave me a cursory tour of the labs, seemed a bit impatient with my questions, and insisted I meet with the company's legal counsel to sign a nondisclosure agreement. We were a long way from the Polonator.

Despite its positive buzz and relatively low overhead, Complete Genomics was hit by the same freight train as nearly every other start-up enterprise in the early part of 2009: extreme investor skittishness and virtually no money to be had. The company cut salaries. Its big ambitions—one thousand genomes in 2009 and twenty thousand in 2010—had to be scaled down in the wake of the global financial meltdown. "Our timing could not have been worse," CEO Cliff Reid told *In Sequence*. "We started this [round of] financing the day that Lehman Brothers failed."[26]

By late summer the markets had rebounded enough to allow the company to close its fourth round: $45 million, which brought its total fundraising to $91 million.[27] Early-access customers had begun queuing up for $20,000 genomes (the $5,000 price tag was still some months away). "For us, sequencing a human genome these days is almost trivial," said Drmanac.[28]

Well then, I said, will Complete Genomics sequence the PGP-10? George had already told me that the company would,[29] but I wanted to hear it from the horse's mouth. "We will do it in our spare time," Drmanac promised.[30]

Twenty-five years in, his love affair with DNA sequencing was still in full bloom. "There is something magical about a complete genome," Rade said right before he rushed off to meet with the VCs. "Being able to say, 'That's all of it!' "[31]

Seven miles up the road, Pacific Biosciences was trying to make its own magic. Founded in 2004, PacBio had gone about its R&D business quietly. But in early 2008, the company's CEO, Hugh Martin, told the *New York Times*, "When we're ready, we're just going to win the X Prize."[32] That same week, at the Marco Island genome technology meeting, PacBio stunned a packed house with a presentation of its Single Molecule Real Time (SMRT) sequencing system. Observers tossed around adjectives like *creative, exciting, thrilling,* and *dramatic.*[33] To close the meeting, the company sponsored fireworks on the Gulf Coast beach. Steve Turner, PacBio cofounder and chief technology officer, predicted that within a few years the company would be able to deliver complete and accurate human genomes in less than an hour.[34]

Turner, however, is hardly the overbearing type. "I'm a huge Steve Turner fan," Chad Nusbaum told me. "Whether the technology succeeds or not, I'm very impressed by him. He's smart, creative, and resourceful, but also quite accessible as a human being. He's not full of himself."[35] When we spoke on the phone I gave Turner every opportunity to take shots at his competitors, but he politely declined. Indeed, he was somewhat reluctant to talk to me at all, if only to honor his more superstitious coworkers. "There is a very famous Silicon Valley flop," he said. "The Segway. Those guys were going to be Google-esque in their success. They hired a biographer who was going to be their official documenter of this historically important invention. But then of course the company failed miserably. Some people here worry we will jinx ourselves by talking."[36]

As a grad student at Cornell, Turner's pal Jonas Korlach had an idea: What if one could actually watch a single molecule of DNA polymerase synthesizing DNA in real time? Turner was already working on nano-structures with the goal of finding better ways to manipulate and visualize DNA. He and Korlach realized that if they succeeded, then they would also be able to "see" which base was incorporated at any given time; that is, they would have the most powerful DNA sequencer ever. But how to bring it to fruition? Their technological breakthrough was something called a zero-mode waveguide, essentially a tiny, glass-bottomed well with metal sides—the whole thing is only a few dozen nanometers wide (about the size of a single virus). When a laser is shone at the ZMW from below, enough light gets in to visualize a single DNA polymerase molecule clutching a single nucleotide (a single "letter" of DNA) but with almost none of the surrounding noise.[37]

SMRT sequencing, if it worked as expected, would be cheaper, lon-ger, and much faster. It would be cheaper because by focusing on single DNA molecules, it would require very little in the way of chemical re-agents to make it go. The machine itself would cost no more than most of the current crop DNA sequencers: probably $500,000–$600,000 (it eventually came online at $695,000). SMRT sequencing reads would be longer because unlike most other methods, the process would not ac-tively terminate enzyme activity in order to build a chain; instead one would simply "feed" the enzyme nucleotides and then let the big dog run. "Our view is that this enzyme [DNA polymerase] is really a sequencing instrument in and of itself, and what a horrible shame to throw it away after every base you sequence," Turner told *Chemical & Engineering News*. "If we free it up to do what nature has programmed it over billions of years of evolution to do, we can get the extraordinary features that it has of extreme frugality and high speed."[38]

By early 2009, PacBio had gotten its average read-length up to 946 bases and shown the ability to produce reads of greater than three thou-sand bases.[39] One of the initial problems with most of the next-genera-tion sequencing technologies—including the Polonator—had been short read lengths: in order to reconstruct a big, complicated genome with the newfangled machines, one had to piece together millions of short fragments against a reference sequence and repeat the sequence many, many times to make sure it was accurate. As I've noted earlier, this was like doing a jigsaw puzzle with millions of tiny pieces, some of which were indistinguishable from each other.* PacBio aimed to increase the size of the pieces, reduce their number, and therefore reduce the difficulty with which they fit together. Eventually the company hoped to generate reads of tens of thousands of bases in length—two to three orders of magnitude more contiguous DNA per read than the current state of the art.[40] Finally, SMRT sequencing would be faster because, left to its own devices, DNA polymerase works fast. In its coming-out paper in *Science*, PacBio showed an average sequencing rate of five bases per second;[41] Turner hoped to multiply that by a factor ten. If he could. And if he could put a million or more wells on a plate (early versions used just a few thousand; the first one was slated to have eighty thousand), then SMRT sequencing could read 100 billion bases an hour: That would mean a complete human genome could be read fifteen times over in fifteen minutes.[42]

At the Cold Spring Harbor Personal Genomes Meeting, with commercial launch presumably less than a year away, Turner was wary of me. When I asked him to describe what PacBio's machine looked like, he said I should look at an article in the next issue of *Forbes*,[43] but conceded that it would only show the instrument with the covers off, not fully assembled.

"If I am considering buying a Tesla," I said, not mentioning that his machine would be more than four times the cost of the electric sports car, "of course I want to see the engine, but, you know, I also want to see the lines."

"I hear you," he said. "Aesthetics are important. But we're not at the right time for that relative to when you'll be able to go to the website and click 'add to cart.'"

---

* This problem became less acute as more genomes were sequenced to higher accuracy that could serve as benchmarks for subsequent genomes.

When I observed that PacBio had raised more than $260 million since 2004 and noted that that was a shitload of money by any standard, his boyish face remained impassive.[44] He shrugged and stuck to the script.

"We are in an all-out sprint before our machine goes commercial. We are following a schedule that has been tightly choreographed."

By the end of the year the company had at least a dozen prototypes and six early-access collaborators. It had begun to assemble a sales force, and it had floated the idea of an initial public offering in 2010.[45] As I poked around, a few (jealous?) competitors harped on the low raw accuracy of PacBio's sequence data; even some would-be customers said, in effect, "Come on already—enough with the fireworks, we want to see human genomes." But the buzz persisted. Asked what sequencing technology they were most excited about, labs responding to an *In Sequence* survey mentioned PacBio most often.[46] The company had had extraordinary success in raising both money and expectations.

George, unsurprisingly, was on the scientific advisory boards of both PacBio and Complete Genomics. "They both have cultures I would love to be in," he said. "They are almost academic places, but they have all the money academics *don't* have. I'm almost ready to quit my day job. They're both front-runners in DNA sequencing and I feel a lot more comfortable being affiliated with both than just one," he said diplomatically. And then he gave me the smile. "Of course it could be there's a third one that wins . . . It could be the Polonator!"[47]

Francis Collins had undergone a metamorphosis.

I first suspected something was up when I watched the webcast of him moderating the ELSI panel at the 2008 Biology of Genomes meeting at Cold Spring Harbor. The subject was Direct-to-Consumer Marketing of Genomic Tests. The panelists were policy expert Kathy Hudson, National Coalition for Health Professional Education in Genetics director Joe McInerney, and, in their first joint public appearance, the Big Three: Kari Stefansson from deCODE, Dietrich Stephan from Navigenics, and Linda

Avey from 23andMe. Faced with an audience of skeptical genome scientists, the company reps were both extending a hand ("we want to work *together* with you," "we *want* to be regulated") and passionate about their model. "I'm actually convinced," said Stefansson in his smooth Icelandic accent, "that in the near future, genetic profiling like [our companies are] marketing is going to power the paradigm shift from interventional medicine to preventive medicine. . . . I think it is always laudable when people learn more about themselves."[48]

After the panelists had spoken, Collins served as interlocutor. His questions were incisive, sincere, and did not betray the anxieties I'd sensed from him in this same auditorium a year earlier. "Are these tests clinically useful?" "Should people be able to get the information whether or not there is a clinical intervention?" "Do we need a database and where should it live?"[49] He seemed to me to be buoyant, open-minded, and fully engaged. I thought maybe this was due to the imminent passage of GINA. But that wasn't the whole story. Three weeks later he announced he would step down as director of NHGRI.[50] He was coy about his plans. The genome community was long on speculation: What was Francis up to?

A few days after he made the announcement I sat on a panel with him and several others at the World Science Festival at New York University.[51] The WSF was organized by physicist/PBS rock star Brian Greene and his wife, TV producer Tracy Day. It featured events all over town, a street fair, films, lectures, and robots shooting hoops.[52] Saturday morning I took the train from my sister-in-law's in Brooklyn to NYU and went to the breakfast buffet for speakers. As I was sitting there with my muffin and fruit cup, an elderly woman approached and asked if she could join me. It was Vera Rubin, perhaps the most decorated female astronomer in history, a former Richard Feynman student,[53] and an all-around lovely person. This was pretty cool.

I went out to the street fair. There I was accosted by self-styled revolutionary communists, who were pimping copies of their pro-evolution, anti-religion tract. I chatted with a scraggly guy who smelled faintly of alcohol. He tried to reassure me that communists didn't think Islam was any worse than any other religion, despite my not having suggested otherwise. "They are all equally bad," he insisted. He continued to press his book on me, but I took my leave—there was a terribly cute animatronic dinosaur I had to see.

Down the street I bumped into friend and fellow panelist Jim Evans, a folksy and perspicacious medical geneticist at UNC and editor of *Genetics in Medicine*. He was curious about consumer genomics, so I had sent him my Navigenics report; he was singularly unimpressed ("Nothing new here. Did this really cost $2500?").[54] I noticed that he was wearing his DNA tie, the same tie I wore at my wedding.

Later I entered the auditorium as Jim and the other participants in our session, "Your Biological Biography," drifted in. Sociologist Nikolas Rose was there. I wanted to tell him what a big fan I was and that I had recommended him for this panel, but thought that might be poor form. Our moderator introduced himself. He was Sir Paul Nurse: London-born president of Rockefeller University, Nobel Prize–winning cell biologist, and extremely funny man. He reminded me of Dudley Moore—short and with the same shaggy mop top. He wore a dinner jacket over a black T-shirt emblazoned with the innards of the human torso. Backstage he told us his own ancestry story: A year earlier he had been detained by homeland security officials because his birth record documentation was somehow lacking in their eyes. He subsequently wrote to the hospital where he was born. His birth certificate arrived and in the space for "Mother" was his sister's name while the space for "Father" was left blank. His presumptive parents were not his parents. He still didn't know who his biological father was. He dryly observed that it "all came back to personal genomics, didn't it?"[55] "I'm not a bad geneticist," he would say later, "but my own rather simple family kept a genetic secret from me for more than half a century."[56]

"Dr. Collins," I said to Francis and extended a hand. He was nothing but cordial. His decision to walk away from NHGRI seemed to have lifted a huge weight from his shoulders. He asked me if I was writing a book. Had he forgotten our tense tête-à-tête the year before? As we waited in the wings to go onstage, I asked him what his plans were. He said he didn't know—he might just enjoy being unemployed for a while. Or he might write his own book about personalized medicine.[57]

He and I would meet again in a few months at the Personal Genomes meeting in Cold Spring Harbor. We interviewed each other over breakfast and this time he let me record him with no preconditions and nothing off the record. Unfettered by NIH (for the moment anyway), he was completely open.

On his current thinking on personal genomics: "We have a long way to go before we can say that all of this is going to provide serious positive-outcome opportunities as opposed to just 'curiosity satisfactions.' But I guess I'm more on the side of, 'Let's go full speed ahead and empower the people interested in having this information sooner rather than later while still making sure we're providing the kind of support they're going to need.' "[58]

Why was he so optimistic? "The most significant reason is that the science has finally gotten to the place where the predictions are not fanciful. Admittedly they are relatively modest risks [in most cases], but they're not made-up. The other thing that's given me an increased sense of confidence is GINA. So we have better science, we're giving out better information, and we have better protections against the most egregious misuses. That really changes the landscape. I think that's why I feel primarily positive versus two or three years ago when I felt primarily negative."[59]

I suggested that, given these developments, the current medical genetics paradigm was not up to the task of dealing with personal genomics and all that it entailed. To my surprise, he agreed. "We know that most common illnesses have heritability that's in the neighborhood of fifty percent. If we had all [of those genetic factors in hand] then the idea that we could just go on with business as usual in medical genetics, which is largely built upon rare Mendelian conditions and chromosomal disorders, is not going to be sustainable. Frankly I'm very worried about my own specialty of medical genetics, a field I'm deeply attached to."[60]

Why has medical genetics not kept up? "We have not succeeded in attracting young, energetic, visionary physician scientists to join us. Here we are on the brink of a revolution where genetics is driving the entire medical profession, yet there are fewer applicants today to medical genetics fellowships than there were twenty years ago! Medical students look around and they don't see a lot of geneticist role models because there are so few of us. We've never been able to catch the wave and build it up. I'm not sure the leaders [in medical genetics], most of whom were trained in an earlier era, are actually all that comfortable with the idea that they need to blow up their profession and start over."[61]

What would you do to fix it? "For starters we really have to renovate the training programs to make them attractive to the best and brightest of the next generation of physician-scientists. [Genetics] ought to be the kind of profession that lots of cerebral young physicians would like to join. Right now we're doing what we've done for the last fifty years: [Trainees] see newborns with birth defects and go to the clinic and see people with [rare] Mendelian conditions. There's not a sense that this is the new genetics—this is the *old* genetics. I think the medical genetics profession [has] more people who are ready to deal with dysmorphology [assessing people with congenital birth defects] than are ready to deal with a genomic sequence that needs to be sorted out. There's still a 'two cultures' problem."[62]

I asked him about "the incidentalome." What happens when people learn potentially worrisome things about their genomes and start to pursue them at great cost and effort only to find out that most are meaningless? Collins said that that was a serious concern. "That's the part I'm most wondering about when it comes to complete genome sequences. For these rare variants it will take decades before we have a sufficient 'encyclopedia' of them to know whether they matter. [In the meantime] we'll be leaving a lot of ambiguity hanging in the air."[63]

We were running out of time before the morning sessions of the meeting would commence. I had caught Collins at a professional inflection point, but his embrace of personal genomics was only the beginning. In the coming months he would establish a foundation meant to address the culture war between science and faith in the United States.[64] He would indeed write a book.[65] In July 2009 he was nominated for the directorship of the NIH.[66] A few weeks later he was confirmed unanimously by the Senate.[67] Had his decision to resign from NHGRI been part of some master plan to become top boss at NIH?

Shortly after his confirmation, he laid out his priorities. First on the list was using hi-tech approaches to discover the genetic bases of diseases.[68] Whatever one's view of Collins, his confirmation was unequivocal proof that genomics had reached the head of the class. And I expected it to stay there: Collins has long been interested in a large cohort study of genes, traits, and environments in the U.S. population akin to the UK Biobank.[69]

There was another, smaller reason I was optimistic that Francis Collins—onetime would-be villain of this book—would do right not only by genomics writ large, but by our flagging health-care system and those who have been ill-served by it. A week after he was nominated to lead NIH I found myself in Washington, D.C., for Genetics Day on the Hill, a day of gentle lobbying of our elected representatives on issues pertaining to genetics or, as was the case in 2009, health care in general. We were given talking points—mostly of the mom-and-apple-pie variety—and we were broken up into teams of three to six people with whom we would visit our senators and congresspeople.

My team included a fifty-something guy from Lawton, Oklahoma, named Dennis Pollock. In 1993 Dennis was diagnosed with alpha-1-antitrypsin deficiency; he had served on the board of the Alpha-1 Association for many years. Alpha-1-antitrypsin is a protein produced mostly in the liver. Its main job is to protect the lungs from an enzyme that digests damaged or aging cells and bacteria. Without alpha-1, the enzyme will attack healthy lung tissue.[70] As a young man Dennis had a double lung transplant. He described himself as completely healthy, although it was clear that schlepping from one congressional office building to another in the July Washington heat took its toll upon him. Being from Oklahoma, Dennis was represented by Republican senator Tom Coburn, aka "Dr. No." Among legislators, Coburn was probably the one most al-lergic to legislation. In 2007 he single-handedly blocked or slowed more than ninety bills, including GINA.[71] Dennis repeatedly came to D.C. to press Coburn about GINA, going so far as to organize a phone blitz on his office. Not long after this—and following a couple of concessions on the part of the bill's sponsor—Coburn lifted his hold on GINA. Dennis's activism won him admirers, including Francis Collins. The two became friends, Dennis told me; Pollock stayed with Collins and his wife when-ever he was in the area.[72] I had heard Collins talk about reducing health disparities on many occasions. It was gratifying to know that it wasn't just talk.

As we collected our trays, Collins asked what I would do with my genome. "What are you going to do when you encounter a SNP no one's ever seen? You will have some *breathtaking* mutations . . . because we all do. You'll be thinking [about what they mean for] yourself, for your kids. What are you going to do with that?"[73]

What indeed.

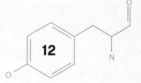

Charity Begins at Home

By the end of the summer of 2009, I was in a pickle. Not only was I on deadline—that is to say, a year late—for this book, but I was getting regular invitations to speak. "O Great Genomeboy, yea, though thou hast walked through the valley of the shadow of death, thou hast feared no evil. We beseech you. Tell us, please: What is it like to have Your Genome Sequenced?"

The truth was, I still had no idea. It had become clear to me that a complete exome, much less a complete genome, was not going to be forthcoming from the Church lab any time soon. While the quality of the Harvard sequence data had gotten better, after two and a half years, I had access to only some 5 percent of my exome, or about 0.002 percent of my genome. This was unlikely to yield the stuff from which deep insights into oneself would be gained. George's lab was an incubator—a place where ideas were born, tested, and published. It was a workshop, not an assembly line. The notion that it would generate thousands of exome sequences like clockwork was simply unrealistic, at least in the near term. It was clear that George needed technical help to bring his grand visions to fruition, but at the moment it was unclear as to when and from whom he would get it outside of his own group. For my own selfish reasons (personal genomics: it's all about me), I needed to look elsewhere.

I began to bug David Goldstein. David is a colleague and friend, and for both relationships I am grateful. He's one of the smartest people I've ever met, and to be honest, I don't have very many really good friends, probably because I have not made time for them, and because I am insecure and afraid that they will reject me. I suspect David doesn't have many close friends, either, though not out of fear of rejection. He is a brilliant geneticist but extremely competitive, sure of himself, convinced that he's right, not one to shy away from a scrap, and not always blessed with the gift of tact. Indeed, he really doesn't "give a toss" (he lived in the United Kingdom for many years and has retained some of the vernacular) about whether you like him. God knows that if he did give a toss, he'd be a mess: over one two-week period, half a dozen people on campus complained to me about him. The general tenor of their complaints was "Who does he think he is?" When I told him of this, he seemed curious about who the aggrieved parties were, but assumed that they were people who, at least on a professional level, did not really matter to him all that much (for the most part he was right).

David grew up in California in a middle-class but broken home, and on the way to fulfilling a passion for marine biology he stumbled into genetics in college. He completed a graduate degree and postdoc with highly esteemed population geneticists, and took a job at University College in London. There he began working in the nascent field of genetic anthropology—that is, using genes to understand where historical populations lived, where they went, whom they mated with, and perhaps something about their culture. A half-Jew who had spent time in Israel when he was young and impressionable, David wondered if genetics could be used to elucidate some aspects of Jewish history. It could. He became a leader of the group that identified a signature (a set of genetic markers) on the

Y chromosome strongly associated with the Cohanim, the ancient Jewish priests of the time of Christ and before, and whose descendants are presumed to carry surnames like Cohen, Cohn, Kahn, Kagan, etc. David and his colleagues also found results that supported the claims of a Bantu tribe (the Lemba) to have Jewish roots. And they used mitochondrial DNA (passed on only by females) to understand how and where Jewish communities were founded.[1] Years later his lab found a signature that correlated perfectly with Ashkenazi Jewish ancestry among people who self-identified as Jewish.[2]

But David had long since grown bored with the pursuit of human history via genetics, feeling that he had taken the science as far as it could go and wanting to do work that was more clinically relevant. He dug into another burgeoning field, pharmacogenetics: the idea that genes have a lot to do with how well (or how poorly) people respond to medications (see chapters 2, 6, and 10). David reasoned that if we could understand the genetic mediators of responses to drugs, then we couldn't help but learn something about the diseases those drugs were used to treat.[3] Particularly in the areas of epilepsy and infectious disease, his hypothesis has turned out to be correct.[4]

He also became an early proponent of genome-wide association studies (GWAS), the approach whereby geneticists collect DNA from thousands of cases with a genetic disease, and thousands of controls without it.[5] By typing, say, a million markers across the genome in hundreds or thousands of cases and controls, one could often find markers that were clearly associated with the trait or disease of interest. Indeed, David and his collaborators (including Duke infectious disease specialist Jacques Fellay and sequencing and genotyping czar Kevin Shianna) used this approach to identify genetic markers that determined how well HIV-infected people handled the virus and the amount of time until they would become sick, if ever.[6] GWAS quickly became au courant and these studies identified markers associated with scores of diseases. The money continued to flow from NIH to fund them.[7]

But because he is honest, because he is often prescient, because he is apt to see the dark side in just about everything, because he is David Gold-stein, by 2008, even as his own GWAS were starting to yield compelling results, he was simultaneously telling the world that the emperor had no clothes and chomping on the hand that fed him. With few exceptions, he said, GWAS was a profound disappointment and he was hereby tendering his resignation from the amen chorus. "There is absolutely no question," he told the *New York Times*, "that for the whole hope of personalized med-icine, the news has been just about as bleak as it could be."[8]

The problem, as geneticists had come to realize, was that we could find variants associated with any given disease, but they didn't explain very much of the disease and so weren't very predictive. "For schizophre-nia and bipolar disorder, we get almost nothing; for type 2 diabetes, [we've found] twenty variants, but they explain only two to three percent of fa-milial clustering, and so on," said David. If schizophrenia was 80 percent genetic, then why couldn't anyone find any genes that played a major role in causing it? Geneticists began wondering where the missing heritability of supposedly genetic diseases could be hiding.[9] Unless it could be found, bringing the genome to the clinic in a meaningful way would be difficult, if not impossible.

That powerful susceptibility genes couldn't be found also explained why so many of the risk numbers reported by the consumer genomics companies needed to be taken with mammoth grains of salt. 23andMe, for example, typed customers for just two SNPs conferring susceptibility to multiple sclerosis.[10] SNPedia, meanwhile, listed twenty-eight of them, all with modest effects.*[11] This was the legacy of GWAS: for complex dis-eases like MS, which are determined by many genes and the environment, any single genetic risk factor was likely to be extraordinarily weak.[12] And even if one had multiple genetic risk factors, as I did for MS, those factors interacted in ways that we were still a long way from understanding well enough to make predictions, let alone help guide treatment.

———————

*   As I fretted in chapter 7, I seem to carry most of these MS variants, though to my knowledge the myelin sheaths covering my nerves are just fine.

David is athletic but neither tall nor imposing. He is thin and some-times socially awkward, almost shy, often looking down with hands in the pockets of his jeans as he makes his way down the hall to a lab meeting, to get a coffee, or to catch a flight to Johannesburg or Taipei. He wears wire-rimmed glasses beneath a mop of thick, unruly hair. His lab is remarkably productive but fairly small given that it oversees an eight-figure annual budget. They are a close-knit bunch, meeting regularly for beer, jogging, meals, trips to the beach, etc. David would not have it any other way. He is free with his money. He rides a motorcycle and smokes cigars. He ap-preciates a funny joke, a well-told story, a nice bottle of wine, a good song. He enjoys life . . . as hopeless and futile as it may be.

David and a few other like-minded people had come to believe that there had to be a better way find the elusive risk factors for inherited traits that GWAS had failed to uncover.[13,14] If, as they suspected, these factors were rare, then they would not be found on the "SNP chips" used in GWAS. So how to find them?

Sequence.[15]

As they continued to chase the genetic basis of infectious disease, David and a group of collaborators amassed samples from a cohort of fifty hemophilia patients who had been exposed to HIV-infected blood but remarkably had not gotten infected. Why not? David assumed there must be one or more genetic variants that made them resistant to the virus. Those variants, he reasoned, would be found by sequencing. Kevin's lab would sequence these fifty people; David's people would analyze the data (when this book went to press, the hemophilia project was ongoing). In the process, Kevin, an Illumina partisan since buying his lab's first Ge-nome Analyzer in 2007, would put all of the company's new toys through their paces.[16]

I saw an opportunity.

"I think you should sequence me," I said to Kevin. "I am already going to be sequenced and phenotyped out the wazoo." (I certainly hoped that that was true.) "My information will be completely public, so there will be zero IRB issues. You won't need to establish a cell line because I'm right down the hall—if you need more blood you know where I live. I am the ultimate control sample . . . Well, unless of course you're studying anxiety, metabolic syndrome, or nearsightedness—then I'm a case."

"Let me talk to David," Kevin said. "It's a question of money."

I pressed the "ultimate control" argument and David began to weaken. "Maybe if you were up for having more of your tissues sampled," he said. "I'll give some thought to it. If I were you I would try to talk to me about it some time when I'm not sober."[17]

"Done and done," I said.

It began with a five-minute blood draw in Clinical Research Coordinator Kristen Linney's office followed by a routine chemical extraction of DNA from my white blood cells. But after that my sample sat for a while. Why? Because the Institutional Review Board at Duke (upon which I happen to serve) didn't quite know what to do with me: most IRBs have not made provisions for returning results to research subjects, even lunatics like me who've *already* seen much of that same data and put it on the Internet for the whole world to examine. It's a situation that IRBs have not ever had to think about—until now. Over the next week the request to have me sequenced and my data returned to me and the world would travel from Kristen's office to the senior chair of the Duke IRB, to the associate dean of research support services, to the director of Duke's Center for Bioethics and Humanities, and then back to the associate dean of research support services, then on to the dean of the medical school and the vice dean for research, then back to the IRB chair, back to Kristen, and finally, back to David, Kevin, and me.[18]

Once we had institutional approval, I assumed we would be off and running. But around that time, the Shianna lab's sequencing runs began to fail at an alarming rate: nearly one in two weren't working. The fluorescent signals emitted by each sequenced base were decaying too rapidly; thus the early cycles could be used from each run but the later ones could not; they were too faint to allow the computer to distinguish one base from another. Within a few days it became clear that it wasn't the machines, but rather the chemicals. And it was not just the Shianna lab. Genome Analyzers all over the world were failing. Illumina, which was forced to eat millions of dollars, told customers it was no longer shipping reagents until it could isolate the problem.[19]

In the meantime, I went back and scrutinized my SNPedia data. Was there anything interesting there that I had missed? I didn't think so. I was seven times more likely to go bald than most men . . . duh. I needed only to look in the mirror to confirm that. I had plenty of risk factors for coronary artery disease, type 2 diabetes, stroke, obesity . . . again, duh. I had had a grandparent on each side die before age sixty from a heart attack. My dad had had a quadruple bypass when he was sixty.

I opened the SNPedia "Medicines" menu with a bit more optimism: as David and others had shown, pharmacogenomics was one of the most promising places for personal genomics to make a difference. Esther Dyson told me that perhaps the most useful thing she had learned from 23andMe was her family's sensitivity to the blood thinner warfarin.[20] As I witnessed firsthand when my nephew Noam developed deep-vein thrombosis (see chapter 10), warfarin is a tricky customer: not enough of it and you might suffer a debilitating blood clot. Too much and you might bleed to death.[21] But unlike warfarin, most of the known drug response markers we knew about did not usually alter one's sensitivity to a clinically meaningful extent.

An exception was a series of markers in the ABCB1 gene. ABCB1 is expressed most strongly in tissues that serve either as barriers (the blood-brain barrier, the placenta) or are involved in eliminating waste from the body (kidney, liver, intestines). One can imagine how this gene might impede drug response: by raising a molecular "gate" or by causing the drug to be flushed from the system more rapidly.[22] But as a team of German researchers found, some versions of ABCB1 genotypes do just the opposite: they *enable* drug response.[23] Perhaps the gene could also lower a gate or keep a drug in the system for an extended period. In any event, most Caucasians, including me, were significantly less likely to respond to certain antidepressants because of their common ABCB1 alleles. I had been on ten milligrams of Lexapro (escitalopram) since 2008 and, as my wife could tell you, it had drastically changed my life for the better, mostly by reducing my anxiety level and keeping my frequent companion, a deep and abiding sense of impending doom, at bay. Yet my ABCB1 alleles would betray me, no?

I went back to SNPedia and looked at the four other non-ABCB1 SNPs that had been associated with depression and for which I had been typed. They were a mixed bag. One was associated with a poorer response to clomipramine,[24] another antidepressant but of a different chemical class than Lexapro, that is, probably irrelevant to me. Another was associated with major depression in Mexican Americans.[25] But 97 percent of Europeans, including me, did not carry the risk allele . . . okay, next. Another was associated with suicidal thoughts in people taking Lexapro's predecessor, citalopram (Celexa).[26] Despite having the "normal" allele, I admit that in my darkest moments I had had such thoughts.

The last "depression SNP," found in a gene that encoded brain-derived neurotrophic factor, was arguably the most interesting. BDNF is an important gene: it's expressed like gangbusters in our brains and is crucial to the way our neurons grow and develop. Knock out both copies in mice and they'll have problems with coordination, balance, hearing, taste, and breathing . . . and they'll die soon after birth.[27] Knock out just one copy and they'll have problems learning and remembering.[28] Like 28 percent of Europeans, I carry one copy of a variant in BDNF that changes a valine to a methionine at position 66 of the protein. Val66Met, as it's known in mutation parlance, has been well studied; however, being well studied doesn't always equate to real understanding. It appears by some accounts to be modestly associated with bipolar disorder,[29] though BDNF has been touted as a "depression gene" in humans for many years despite inconsistent evidence and frequent failures to replicate. Some psychiatric geneticists think it's time to move on.[30] But we know that Val66Met has functional consequences. People with the valine allele exhibit higher BDNF activity; they tend to perform better on memory tasks. And the anatomy of their brains is different than people who are Met/Met.[31] As I wrote this, pharmacogenomic data were scanty, but some studies suggested that people with lower levels of BDNF responded better to antidepressants.[32]

Without knowing my BDNF levels, I knew this was all speculation. But as far as BDNF being a common "depression gene," David Goldstein was having none of it. "Until fairly recently, psychiatric geneticists would start out believing in a gene because a role for it made such good sense, and then they would go out and find themselves a polymorphism in the gene that showed an association with *something*: a cognitive trait, an imaging trait, or with some disease of interest. That the association statistics were weak was largely judged acceptable because it all made such 'good sense.' Of the many famous genes implicated in this era, none survived into the genome-wide phase when researchers cleaned up their acts and insisted upon consistent standards of evidence. Even career-creating 'discoveries' like polymorphisms in BDNF associated with cognitive performance have failed to find support in the much better powered and better controlled genome-wide studies using large samples." Both he and our colleague Anna Need pointed out, however, that truly *rare* variants in BDNF might yet turn out to be responsible for some subset of psychiatric illness.[33]

This never-ending exploration of my own cellular White Pages was giving me a headache. I turned my attention back to sequencing. Eventually a millionth of a gram of my DNA would make its way into Kevin's lab and into the pipeline, unstable reagents be damned. Kevin warned me that because of the reagent problems, there was a fair chance my sequencing runs would fail; we would know in the next day or so.

Inside the large windowless room that housed the sequencers, the hum was constant. Internal fans kept the machines cool. Every piece of equipment seemed to have its own computer. The first step was for my DNA to be fractionated by a vibrating machine called a nebulizer into super tiny bits some 200–400 base pairs long—the DNA needed to be small before it could be amplified en masse in a later step. If the pieces were too big, they would take up too much room on the flow cell, the small glass slide where the business of DNA sequencing would get done. The ends of the DNA were "polished" enzymatically to make them uniform, and two unique adapters (short stretches of synthetic DNA) were attached to the ends with another enzyme (ligase) to the fragments. The adapters would allow the enzyme to recognize every fragment equally and sequence it. The fragments were then amplified by PCR (see chapters 5 and 9) to make sure there were sufficient ligated fragments for the actual sequencing reaction still to come. Once this step was done, the sample preparation steps were over: a "library" of my genome—millions of fragments with adapters attached—was ready. "Everything to this point is pretty straightforward," Kevin said. "Illumina produces a kit for these steps and it's basically idiotproof."[34]

The surface of the flow cell was a tiny grid the size of a microscope slide: eight lanes and one hundred microscopic tiles (think ceiling tiles) per lane. The goal was to get about two hundred thousand amplified "clusters" of DNA per tile. But we're getting ahead of ourselves. The next step was to coat the surface of the flow cell with a dense lawn of primers (again, short stretches of single-stranded DNA) that were "flapping in the breeze." These were complementary to the adapters attached to the fragments of my DNA: they would recognize each other and stick together, or hybridize. They would do exactly what their name suggested: prime the reaction by giving the enzyme a known sequence to start with. But first we separated the strands of my DNA with sodium hydroxide; making the clusters single-stranded meant they would bind to the primers on the flow cell. Once the single-stranded, adapter-ligated fragments were bound to the primers on the flow cell, they could be amplified again to scale up the reaction. This step was called bridge amplification because each end of a ligated fragment would form the middle of a bridge with one of the primers on the surface of the flow cell. At the end of this process, which was done on a separate bench away from the sequencing machines at a so-called cluster station, each fragment was amplified about a thousandfold.

The DNA was also double-stranded for the moment; DNA is more stable in this state and can be kept in the fridge for a couple of months.[35]

As I watched Kevin's technicians, Ryan Campbell and Linda Hong, go through the steps of sample preparation and bridge amplification, it struck me as an intense, monastic exercise: it demanded concentration and meticulousness, but was also repetitive and trance-inducing.

The next day we were ready to sequence. The lab techs made the DNA single-stranded again so that they could attach a primer specifically for sequencing. Campbell, a tall and shaggy guy, opened the blue Plexiglas door to the Genome Analyzer IIx, itself a big blue box. He complained about how difficult it was to get his hands into the machine to load the flow cell. He checked to see if the reagent tubes were pumping—the presence of bubbles meant the chemicals weren't smoothly flowing in and out of the GAIIx. Inside were a glass prism and a camera that would image my DNA. The sequencer would take test photos and use that information to allow the user to focus the camera in three dimensions. "If it's not focused, you're screwed," said Kevin. "Once it is, then you can go to town: you can start the sequencing cycles." The first cycle of sequencing would incorporate a single fluorescent nucleotide in billions of reactions across the flow cell, followed by high-resolution imaging of the entire flow cell—this was what was meant by massively parallel sequencing. After the first cycle we looked at how it was going: the computer attached to the sequencer reported that we were getting about 180,000 clusters per tile—not bad. The machine would now repeat this seventy-five more times, one base at a time, generating a series of images with each one showing a single-base extension at a specific cluster. On the screen the images of clusters looked exactly like the images of polonies I had seen on the jerry-rigged computers in George's lab in 2006: a starry night. After seventy-five cycles, the Shianna team would repeat the process another seventy-five times, starting this time from the *other* end of each DNA molecule. So-called paired end sequencing was more accurate and had become an industry standard. This process would take about three days.[36]

After that, the sequencing was done; now it was up to three computer programs named after birds to read it. The first, Firecrest, converted the images into numbers. The second, Bustard, took each number and assigned a base—A, G, T, or C—to it. The third, Gerald, filtered the results and aligned them to the reference sequence put together by the Human Genome Project. Sequence deemed to be of high quality was kept; low-quality data were thrown away. Another program, the Burrows-Wheeler Aligner, would carry out more refined sequence alignments.[37]

The Shianna lab managed to get about 80 billion of my base pairs at high quality: mine would be a 26x genome (80 billion base pairs ÷ 3 billion base pairs per genome = a genome that has been read 26 times over). When one filtered out duplicate PCR reactions and other meaningless genomic detritus, it was about 23x. Not the shiniest genome ever sequenced and certainly a level of quality that would be scoffed at before too long, but pretty good given the 2009 state of the art. Generating my sequence at 25-fold redundancy took about a week from start to finish. "The GAIIx is so easy," said Kevin. "Suddenly we're a genome center, just like that."[38]

A few days later I went to the lab and pulled up a chair. Reggae pulsated in the background along with the white noise of the sequencers' fans. I was there to discuss the gestalt of my sequencing runs with lead technician Jason Smith. He came to the Shianna lab in 2006; he was Kevin's first employee. He had a ruffle of close-cropped, dark blond hair, glasses, and a beard. The serious academic look belied an easygoing nature, although he spoke faster than any other southerner I've ever met. Raised in Greenville, North Carolina, he had studied biology at North Carolina State. When I asked him how he became a next-generation sequencing guy, he shrugged and said it was mostly by chance. "I enjoy working with computers."[39]

I brought along the spreadsheet Jason had emailed to me, a blur of numbers and acronyms. "I always look at the error rate first," he said. "Generally the most we can tolerate is two percent. Yours is a little high—right at two percent—because these defective reagents are causing a faster decay of the intensity." He explained that the 2 percent figure was itself inflated because it included legitimate SNPs that, until they could be identified, would appear to the computer to be errors. Because Illumina read lengths were still a fairly short seventy-five base pairs, aligning them to a reference genome was trickier than with old-fashioned Sanger sequencing and its 800-base-pair reads, especially if there were too many errors (recall the jigsaw puzzle analogy). When I was sequenced, Illumina's specs called for a 1.5 percent error rate or better.[40]

Like most people, I had about 3–4 million SNPs, so the true error rate—bases that were actually "wrong" and not actual variants in my genome—would be well below 1 percent. And, like most everyone else, about 80 percent of my reads aligned to the reference sequence, that is, the composite genome that came out of the Human Genome Project circa 2003 and has been periodically tweaked and updated ever since.*[41]

What about the other 20 percent that didn't align? I wondered. Is one-fifth of our DNA from extraterrestrials? Jason said the unaligned sequence was a combination of any number of things: true sequencing errors, uncalled bases the software couldn't recognize (labeled simply "N"), or multiple SNPs in the 75-base-pair chunks spit out by the machine and which, again, the computer would think were errors and be reluctant to align with the reference sequence.[42]

The software would reject bases it couldn't read based on their low intensities. Such bases were said to have "Failed Chastity," Chastity being the whimsical name of the filtering program. The remaining "chaste" data were now ready for SNP calling, the process of distinguishing errors from real variants.

In the lab Kevin asked me to hit the delete key on the computer that sat next to the machine that had sequenced my genome. The computer informed me that the 1.1-terabyte file was too big for the recycle bin and did I *really* want to permanently delete it? I said yes and within a minute it was gone. "Congratulations, you just deleted twelve thousand dollars," said Kevin. "And a week's worth of work."[43]

---

* As mentioned in chapter 2, much of that original genome probably belongs to a guy in Buffalo code-named RP11, who unwittingly became a big part of the first reference genome.

I moved into the interpretation realm down the hall. When I knocked on Dongliang Ge's doorjamb he immediately put down his lunch and jumped out of his chair uttering apologies. He was short and well groomed: black hair parted on one side, pressed khakis, and a button-down shirt. He was good-natured, perhaps a little high-strung, and spoke with a bit of a stutter, which probably had something to do with English not being his first language. He trained as a biostatistician and genetic epidemiologist in Beijing. On his Duke faculty website was a quote from Albert Einstein: "Out of clutter, find simplicity. From discord, find harmony."[44]

His crowning achievement, at least in his young career thus far, was a tool designed to extract the simplicity and harmony from human genomes. He had developed a suite of Java-based software tools that provided a user-friendly interface to annotate, visualize, and help interpret the reams of data emerging from whole-genome association and sequencing studies.[45] These programs were so useful to the lab that they had "saved our asses," according to David. Dongliang began to tutor me on the use of the Sequence Variant Analyzer, which let one look at the genome at any level of resolution: from an individual SNP to an entire set of twenty-three chromosomes.[46]

"I want to know what are the genes related to the SNPs in your genome," he explained. "What is the ontology? What are the pathways? We can't get that from text files of sequence. We want to know the likely *functional* impact of variants in your protein-coding sequences. An amino acid change may not matter at all or it could be of great biological consequence."[47]

He gave me the caveats. "Of course, you are but a single, healthy individual. We can't do real statistical testing just on you. We can look for places where genes appear to stop prematurely. We can look at large structural rearrangements of your DNA. But we will probably not know what is causal."[48]

"That's okay," I told Dongliang. I had long since consented to uncertainty.

"Let's load your genome," he said. Even he seemed a little excited; I was the first sequenced genome he knew personally. Before we could do that, however, we had to dial up the warp drive. Dongliang pressed the button on the CPU under his desk and a loud whirring noise could be heard. His PC was essentially a server: it had forty-eight gigabytes of RAM and a massive hard drive.[49] In time, everyone in the Goldstein lab would have this type of computing power, which, by the time you read this, might not seem like such a big deal.

Dongliang explained to me that we could look at three files, each of which we would then filter so as to cull my genome of noise and things we were unlikely to be interested in, that is, the vast majority of the genome. The first file would contain all of my 3.6 million SNPs. The second would list the approximately 700,000 insertions and deletions ("indels") that were less than 75 base pairs, or about the length of a single read on the Illumina sequencer; any insertions and deletions longer than that could not be observed directly—they could only be "predicted" with lower confidence. These larger, less certain disruptions were in the third file. Another file harbored all of the 26+ sequencing reads in total, but as I found out rather quickly, it was simply too large and overwhelming to browse unless you knew exactly what you were looking for . . . or had way too much time on your hands.

The loading took about ten minutes. "We will first perform quality control on your SNPs. We have a consensus score that reflects a number of parameters, including how well the short reads can be aligned to the reference and how confidently we can call SNPs." A bar graph popped up on the screen. "Your curve is the typical normal curve," he said. "That's good."[50]

"How does the program know what's a real SNP versus what's an error?" I asked.

"If we see it in at least three separate reads, then we are confident it's not a mistake," said Dongliang. This criterion immediately got rid of four hundred thousand false SNPs. He did the same for small indels. "Now we can explore," he said. On the screen appeared an animated ideogram of the X chromosome: a horizontal oblong pinched slightly off-center (the centromere, where chromosomes pair with each other) and was marked with black and gray irregular vertical stripes representing the way chromosomes take up stain under a microscope.[51]

He zoomed in and pulled up the Factor VIII gene. Since he had been working with hemophilia samples, he was familiar with this gene: it makes an essential blood-clotting protein, and mutations in it cause classic hemophilia.[52] He studied the different-colored vertical dashes that represented variations along the rectangular diagram of the gene. "Well, you have nothing in any of the protein-coding exons for Factor VIII. Nor do you have any functional indels. That's good," he said matter-of-factly, "because otherwise you would be expected to have hemophilia."[53]

I imagined old-school clinical geneticists and genetic counselors cringing at the casual, cavalier nature of the discussion we were having. What if Dongliang had seen something unusual? Had he been melodramatic, he might well have said, "Oh my God! You'd better go see a hematologist . . . stat!" The Watson and Venter genomes had been curiosities—privileged men who had played major roles in the elucidation of the human genome. Maybe there was a sense among some that they were somehow *entitled* to their genomes and deserved to be first in line. But now, with people like me and the PGP-100 on the horizon, the hordes of mostly healthy barbarians were starting to arrive.

"Now let's compare your Factor VIII to one of the hemophilia genomes," Dongliang said. "This individual has a fourteen-base-pair indel in a coding exon. This is a very bad thing." The hemophilia patient's Factor VIII gene was shaded blue and black like mine . . . until two-thirds along the length of the gene; there the Sequence Variant Analyzer had shaded it green. "Green is everything after the indel," said Dongliang, "which causes a premature stop."[54]

Our genetic code operates in triplets called codons: the consecutive bases AAA code for the amino acid lysine. TAC codes for tyrosine. AGG codes for arginine, and so on—there are sixty-four different triplets (three-letter combinations of A, C, T, and G), sixty-one of which code for the twenty amino acids used to make a human (the code is redundant, so most amino acids are coded for by more than one codon). Three of the triplets, however, do not code for an amino acid. They are called stop codons. They signal the end of the coding part of the gene and therefore, the protein. You can imagine codons I mentioned above lined up consecutively in a gene: AAA TAC AGG. In the corresponding protein (the average protein is about four hundred amino acids long), one would see the chain of amino acids lysine-tyrosine-arginine resulting from these three codons. Now imagine that a cellular accident occurred and the three bases before the last one were deleted. In the gene we would now have: AAA TA - - - G, with the dashes representing the deleted bases. As the cellular machinery transcribed this DNA sequence into RNA, which would then serve as the template to make the protein coded for by this gene, it would read the mistaken sequence that had suffered the deletion: AAA TAG. The first amino acid, lysine (AAA), would be read just fine. But the codon "TAG" does not code for an amino acid; it is a stop codon. The protein would simply end there. This is exactly what happened in the case Dongliang showed me.

"This individual has lost all of his Factor VIII exons after exon fourteen. That causes hemophilia A. I think your genome, on the other hand, is normal," he said. "At least for Factor VIII."[55]

My lesson in genome navigation was over. Dongliang installed a graphical user interface on my laptop so that I could peruse my genome from anywhere.

I was on my own. I went to my office and closed the door.

And immediately opened it again. Trying to get the server to run from my laptop was hopeless—it was molasses-slow and kept locking up. I would have to find a desktop with lots more memory and a giant hard drive.

I was only beginning to appreciate the idea that 80 billion bases was, undeniably, a metric *shitload* of data. Setting filters and plucking out what you were interested in was one thing, but even just moving the data around, waiting for the program to load, and making sure that the particular nucleotide you were looking at was where it was supposed to be and not in the half a percent that was actually wrong, was painstaking. Hugh Rienhoff called it "hand-to-hand" combat.[56] I was about to earn my stripes, or at least get a good bayoneting.

Another problem was that genome interpretation software was still not designed for civilians. In part this was because the data weren't meant to be parsed by civilians. They were for bioinformaticians and various other übernerds: people who were used to walking and talking in Linux and writing code on the fly. If a particular analysis function didn't work, they could intuit a workaround. I most certainly could not. Over the next several days I could be found in Dongliang's office pestering him about how to get the Sequence Variant Analyzer to do what I wanted it to do in a way that did not tax my nonexistent computational skills. He was too nice to tell me to get lost.

If you're above a certain age, you'll remember the early days of the Internet: few Web pages, dial-up access, frozen computer screens, and frequent rebooting. Good times. In the realm of genome interpretation circa 2009–2010, we were still in dial-up mode. SNPedia, for example, had been online for less than three years and had annotated less than 12,000 of all of the millions of validated human SNPs[57] described in dbSNP, the NIH's central repository of SNPs and other small variants.[58] Despite this, SNPedia and its creators had already become a critical resource for SNP annotation: the personal genomics companies used it, the PGP used it, and hobbyists used it.[59]

Meanwhile George's group had developed its own tool to make clinical sense of genomes, Trait-o-matic,[60] an open-source (natch!) program that linked out to SNPedia, Online Mendelian Inheritance in Man (OMIM), the Pharmacogenomics Knowledge Base,[61] and, pending commercial considerations,[62] the partially user-funded Human Gene Mutation Database.[63] But within those four databases' purviews—SNPs, genetic diseases, genetic markers that influenced drug response, and all human mutations, respectively—there was not much in the way of homegrown observations. SNPedia would rank your SNPs, but in a fairly subjective and sometimes arbitrary manner determined by the community. When I asked Mike Cariaso why SNPedia had ranked my elevated risk for male-pattern baldness as my most interesting trait when I was at higher risk for heart attack, diabetes, and God knows what other serious medical condition, he wrote back with a shrug: "You're free to change it to whatever you think is more appropriate."[64]

The onus was, finally, on the user to prioritize his or her variants. 23andMe and Navigenics made choices about what to show customers, I surmised, based on what they viewed as clinically significant, interesting, marketable, "actionable," and what markers happened to already be on their standard SNP chips. For both companies, the information to be returned amounted to SNPs predisposing to a few dozen traits, which made sense. The expression "drinking from a fire hose" was apt when it came to genomic information: if the direct-to-consumer companies overwhelmed their customers with thousands of genotypes and each one was perceived to be a potential time bomb (or perhaps just as bad, a time *waster*), this would not have been a wise business strategy. Thus I wanted to know how Trait-o-matic prioritized my variants before I got sucked into another vortex of databases, articles, and spreadsheets.

George's team, including geneticist Joe Thakuria, had outlined this in a paper they had just submitted. Because some 90 percent of disease-causing mutations occur in protein-coding regions, they chose to focus on those in particular, and especially on those already used in genetic testing. And because the genomes (or partial genomes) came from healthy people, that is, the PGP-10 and fifteen others, they didn't expect to find many strong ("highly penetrant") mutations. And they didn't: just eleven in all. But both OMIM[65] and the database of genetic testing[66] listed just a small minority of even *potentially* harmful variants in the human genome. This made evolutionary sense: if clinically important genes were mutated all the time, then we wouldn't be here to talk about them because we would have died in utero. The genes associated with serious diseases were telling us something: they were telling us that they mattered.[67]

To identify changes in those genes and others that might be of clinical interest, the Church crew first generated a list of all of the variants they found in the twenty-five genomes they studied (some were incomplete). They matched those lists to the more than 1,500 variants they found in SNPedia and the Pharmacogenomics Knowledge Base. They then looked for those variants that resulted in an amino acid change: those would presumably have some effect on the protein and perhaps raise one's risk for a single-gene disorder such as, say, ALS. SNPedia was also useful in finding genes that contributed to complex diseases, even if those genes did not cause disease outright. As I've mentioned, I carried a change in a gene that raised my risk for rheumatoid arthritis fivefold above average, for example, though it was hardly the only rheumatoid arthritis susceptibility gene we know about.[68] In the context of all of those RA susceptibility genes, my risk was probably lower than five-fold above average, though a rheumatologist assured me that no one really knew.

The Church lab also looked at changes that were known to disrupt splice sites. On their journey from DNA to protein, most genes are spliced: after the DNA is transcribed into RNA, it is cut into pieces and the protein-coding segments, the exons, are rejoined to each other and serve as the template for the protein while the bits that have been excised, the introns, are discarded. Thus the post-splicing version of an RNA molecule might be only a fraction of the length of the genomic DNA that gave rise to it. The cool thing is that the same gene can be spliced in multiple ways—different splice forms can be used in different tissues to produce similar but not identical proteins. Splicing is therefore a terrific source of protein diversity. Of course, one can imagine that if the sequence that instructs the cell to splice it goes awry, the cell won't get the right message and bad things could happen. Aberrant splicing has been shown to play a role in certain instances of many diseases, including retinitis pigmentosa, muscular dystrophy, breast cancer, lupus, dysautonomia, cystic fibrosis, and elevated cholesterol.[69] Both Sequence Variant Analyzer and Trait-o-matic looked for these types of mutations, too.

In general, for a variant to raise a red flag, it had to be:

- rare (if it was common, it would be less likely to cause severe disease or we'd all be sick or dead)
- clearly associated with disease (false positives have been the bane of geneticists for decades)
- likely to actually cause an observable phenotype (you can't measure it if you don't know it's there)
- shown to be clinically important in the literature (new occurrences were always harder to prove)

Making a "final" determination of clinical relevance still had to be done by hand, at least for now. Of the variants discovered in the PGP-10, only one was deemed to be serious. This was found in Steve Pinker (PGP6), who carried a mutation in the MYL2 gene, which had been shown, in some cases, to cause hypertrophic cardiomyopathy (HCM), a thickening of the heart muscle that makes it harder for the heart to pump blood. HCM is among the major causes of death in young athletes during strenuous exercise.[70] When I asked Steve about it by email, he said it was certainly a surprising discovery, but he seemed pretty sanguine. "I had some mildly anxious thoughts between the time that Joe Thakuria went over the pedigrees with me (which showed that the association was real enough to follow up, but too tenuous to get upset about) and an echocardiogram which revealed I am fine."[71]

I hoped that I would be as serene. I would find out soon: I was finally ready to turn my attention to my own genome.

13

Antarctica

My wife, the lapsed Catholic, teaches Jewish preschool and Sunday school and by now probably knows more than I do about Tisha B'Av and kosher dietary laws; she plays a mean "Hinei Mah Tov" on the guitar. I, on the other hand, since the day I donned a powder blue leisure suit in 1977 and read from the Torah on the occasion of my bar mitzvah, have spent most of my life—much to my parents' chagrin—as a "twice-a-year Jew."* I had a lengthy flirtation with Israel in the 1980s, and even lived there for a year, but I was unwilling to make the full Zionist commitment—something about the prospect of Katyusha rockets falling from the sky and taking up arms just didn't work for me. I am prone to bouts of self-hatred, but I've never denied my heritage. So it struck me as significant somehow that at the beginning of my foray into personal genomics in 2006 I spent an intense hour with Rabbi Terry Bard, a pastoral

---

* Yom Kippur and Passover. Lately I'm down to once a year—sorry, Mom.

counselor at Beth Israel Deaconess Medical Center (see chapter 2), talking about the PGP, George Church, informed consent, and Jewish notions of free will; and at the end of my journey, in 2009–2010, when the time came to really get a handle on what, if anything, I should care about in my genome, I spent an intense hour—and had an ongoing correspondence— with George's then–grad student Abraham Rosenbaum, an observant Jew from New York with close-cropped hair and wire-rimmed glasses who is warm and generous and talks very fast. I would frequently email him, ask for a data file or a link as well as a layman's explanation of what it was I was asking for, and he would write back a detailed response that usually began with something like "No problem. If it were not for people like you then I would have nothing to work with."[1]

Abraham explained that the PGP's prior sequencing failures were not really the Polonator's fault—the Polonator, of course, had not yet made much of a dent in the game. Instead our initial data were supplied by the core sequencing facility that served the Harvard Genetics Department with Illumina machines. Abraham had taken anything that the Trait-o-matic had flagged as suspicious and then gone back and resequenced those variants using good old-fashioned Sanger sequencing, which was still the gold standard for quality. "So," I wondered aloud, "when can I see these data?" By now this question, asked frequently of George, had become both comical and rhetorical; my expectations were low. To my surprise, however, Abraham offered to let me look at the latest version of my own Trait-o-matic report. I was stunned. It didn't seem possible. At long last, I would see actual PGP sequence data from my own genome![2]

I logged on to look at it and was greeted by a warning I knew all too well:

> Before using Trait-o-matic, users should be aware of ways
> in which knowledge of their genome and phenotype could
> be used against them. For example, in principle, anyone
> with sufficient knowledge could take a user's genome or
> open medical records and use them to:

1. infer paternity or other genealogical features;
2. claim statistical evidence that could affect employment or insurance;
3. claim relatedness to infamous criminals;
4. plant incriminating synthetic DNA at a crime scene;
5. reveal susceptibility to diseases currently lacking a cure.

Furthermore, any genetic information obtained about an individual may also have relevance to family members.[3]

By now I was not losing sleep over any of this, but I was still curious about the last bit: Would there be actual clinical relevance for my family?

Most of what the Trait-o-matic generated was stuff I knew about (increased risks for diabetes, lupus, bipolar disorder), had never heard of and couldn't pronounce (molybdenum cofactor deficiency; alpha-methylacyl-CoA racemase deficiency), and/or couldn't get too worked up about. One variant interpretation, for example, suggested I "may require more methadone during heroin withdrawal." Duly noted. Eat your heart out, Keith Richards.

Among the most potentially interesting variants were those in OCA2, an oculocutaneous albinism gene. OCA2 is associated with albinism and certain general pigmentation traits such as eye color, skin color, and hair color. I carried two variants in this gene: A481T and R305W. R305W makes one more likely to have brown or black eyes—my eyes are brown, so again: woo hoo! The other variant, A481T, seemed to be fairly common in Japanese people. But it wasn't clear if it actually caused albinism; if it did, it probably needed help from at least one other mutation.[4] I was not albino, though I am blind as a bat without my glasses. So why did I care? My nephew Jesse, in addition to having Hirschsprung's disease (see chapter 1), had albinism: he had light brown hair, fair skin, and was myopic—he began wearing glasses as a baby. Was there a connection? I didn't know but I wondered. I emailed my brother and sister-in-law, who graciously offered to indulge my curiosity and said they would ask the eye doctor for Jesse's report. They didn't seem to share my interest in Jesse's genomic underpinnings and I couldn't blame them. They had gone through three years of hell and weren't about to start drawing still more blood from their son just to satisfy his uncle's curiosity. And I knew it probably didn't matter. Jesse was in good health overall and already had what Hugh Rienhoff would call a "sound management plan."[5] Jesse's parents used sunscreen on him and he got regular checkups at the eye doctor. His molecular defect had an invasive but effective solution. A link between Hirschsprung's disease and OCA2, or even Jesse's particular case and OCA2, was not obvious. Yes, some of the same ancestral cells that go on to populate the gut also go on to populate the skin and involve some of the same biochemical pathways,[6] so perhaps OCA2 had something to do with Jesse's phenotype. But again, it was of academic interest only to me and perhaps a few developmental biologists somewhere.

When I got my complete genome from the Goldstein lab, one of the first things I did was to check my genotypes in the RET gene, the major Hirschsprung's susceptibility gene I had worked on. They were completely normal. And even if they hadn't been, what conclusion could I have drawn, other than that God had a twisted sense of humor? Jesse Angrist was not Bea Rienhoff. He did not need his father's—let alone his uncle's—heroics. Most of us were fortunate that we didn't.

The other "rare" change turned up by the PGP that was connected to disease was in one of the dozen or so genes responsible for Fanconi anemia, an autosomal recessive disorder characterized by short stature, skeletal defects, a higher incidence of cancer, bone marrow failure, and cellular sensitivity to chemicals known to damage DNA.[7] Jim Watson, Rosalynn Gill-Garrison, and the first Asian genome (Chinese) carried it as well. Since it was recessive and I was done having kids, it was not a major cause for concern (I don't have Fanconi anemia, although I am short). Of course, it was entirely possible that one or both of my daughters carried it; I considered that to be something worth remembering.

And what about my own kids? I had gotten good news from the breast cancer mutation test: I did not carry any of the three most common mutations found in Ashkenazim. But BRCA1 and 2 were big genes. More than 1,600 distinct variants had been reported in BRCA1[8] and more than 1,800 had been reported in BRCA2.[9] In each case more than half of them had been seen only once. Rare was usually good when it came to cancer, but of course rare was also what so often went undetected. I was a bit taken aback to see a fairly rare variant in one copy of my BRCA1 gene. At amino acid position 1564, I appeared to carry a change that altered a histidine to a proline, a "nonconservative" change from a basic amino acid to a neutral one. I became a bit agitated: After my DNA Direct test for the Ashkenazi mutations, I had more or less dispensed with worrying about being a carrier of a breast cancer mutation. Had I exhaled too soon? Right away I began digging into the breast cancer genetics literature but couldn't find much. In part this was because what the Trait-o-matic had shown me included a typo: the actual mutation led to a *leucine*-to-proline change in the BRCA1 protein, not a *histidine*-to-proline one (not that that was necessarily any better or worse). Once I figured that out, I went looking for papers that had analyzed L1564P instead of H1564P. In a study of African American women, L1564P had been seen in patients with breast cancer but not controls.[10] This was interesting, but not authoritative: the sample was small and the mutation detection method used in that study was not perfect. There was nothing to say this wasn't an innocuous variant. Or a death sentence. Heh. "Absence of evidence is not evidence of absence."

I needed a higher power and I found one: the Breast Cancer Information Core (BIC), a consortium hosted by NHGRI that had taken upon itself the job of determining the clinical significance of every single one of the thousands of variants observed in the breast cancer susceptibility genes. I filled out the online application, gave my Duke credentials/affiliations, and assured BIC that I was only doing "research." Once inside I easily called up L1564P and found a wealth of information about it: how many times it had been seen (twelve), in what populations (mostly African), what did this type of change mean for the BRCA1 protein (most likely nothing), and the bottom line: "Overwhelming evidence from sequence conservation, epidemiologic studies, [and] co-occurrence with different deleterious mutations [suggests] that this variant is not a signifi-cant cancer susceptibility allele." Woo hoo!

BIC turned out to be a wonderful resource. It epitomized what a ge-nomic database should be: it housed all of the necessary and sometimes inscrutable reams of population genetic data, but it also showed what could be done when information was not only aggregated but interpreted with an eye toward actual human beings' health. Of course, BIC cau-tioned against using its information for clinical purposes and despite being labeled "open-access," BIC was available only to "qualified investigators." We can't have civilians poking around in these minefields, now can we?

In my whole-genome data, I found another change in the protein-coding portion of BRCA2 and emailed breast cancer geneticist extraor-dinaire Mary-Claire King again. She didn't remember me and probably thought I was yet another kook bugging her for wisdom (which I was, but never mind). She kindly informed me that the variant I carried was seen with equal frequency in both breast cancer cases and controls, that is, almost certainly nothing to worry about.[11]

But mutations don't happen only in protein coding regions. If they occur in introns (the discarded parts), they can still mess up the transcrip-tion process that gets DNA to RNA. And so I went exon by exon (BRCA1 has 24 exons; BRCA2 has 27) and intron by intron and took a look at each of the eighty-six SNPs and twenty-two indels I carried. I could ignore many out of hand because they had already been seen in other databases full of healthy people. As for the rest of the variants, the sequence and SNP databases—even BIC—often weren't much help. The clinical sig-nificance of these changes, their frequency in the population, and even whether they were passed down in predictable Mendelian fashion—all of these facts were often unknown. It turned out to be a couple of hours of work that was somewhat reassuring—the chance I carried a "non-Ashkenazi" mutation was pretty small to begin with. But even this was not an unconditional guarantee. And frankly, even for a genome geek the whole business got to be tedious. I was ready for the dial-up age to be over.

Thanks to Dongliang's patience, I was finally able to download selected other subsets of data. One was a file of all of the variants in protein-coding genes that inserted or deleted bases and disrupted the proper reading of amino acids—so-called frameshift mutations. In 148 instances I carried two copies—one from each parent—of these types of mutations. In twenty-seven of those, the frameshift introduced a premature stop codon as it did in the Factor VIII example Dongliang and I had looked at; presumably premature-stop mutations would yield no protein. Francis Collins had warned me a year earlier: "You will have some *breathtaking* mutations." The thinking was that if one had zero functional copies of something instead of one or two, then that might be expected to have an effect. Or not: I found that I carried two defective copies of the CASP12 gene, which is an important gene involved in inflammation. But there's a good chance that you do, too. Most humans carry mutated versions of CASP12 and produce a truncated, nonfunctional protein. The genome often makes do with less. And perhaps it's not just making do: nonfunctional versions of genes like CASP12 may actually be a selective advantage. This might help to explain why each of us walks around with dozens and dozens of broken genes.[12]

I kept on in hand-to-hand-combat mode, going gene by gene. And time and again, the paucity of information was striking: I would find mutations in genes that coded for proteins, but the proteins' ascribed functions would be so general and/or tentative ("may be involved in transcriptional regulation") as to be meaningless. In some cases, the proteins didn't even have names, let alone functions assigned to them.

Another striking realization was that our genomes are an utter mess: chunks of DNA are not only missing, but they have often moved to different chromosomes, or flipped around. Genes we don't need—ones that code for olfactory receptors, for example—may be completely absent. Why don't we need them? Because we have hundreds more where those came from, for one thing. And in evolutionary terms, other than for detection of the ocasional noxious fume, we don't necessarily need a sense of smell to survive anymore. The aroma of food can be pleasurable (pizza, popcorn, fresh-baked bread, curry) or repugnant (rotten fish, rancid Limburger cheese, your brother after he's eaten a burrito), but these days Westerners generally don't depend on it for hunting and gathering.

The genome was also a source of pure banality. We all have "house-keeping genes" that are turned on all the time at the same level, doing the same boring job in nearly every cell in nearly every tissue. Like olfactory receptors, these genes are often redundant; one fewer copy would prob-ably not be missed.

And some genes appear to be present but not real: they are pseudo-genes. They typically have many if not all of the hallmarks of a gene: a beginning, middle, and end; a plausible sequence of amino acids; a regu-latory region; binding sites for proteins. But these "genes" don't lead to RNA and therefore they don't lead to protein. There may be every reason to believe that the proteins encoded by them should exist, but no one has ever seen them.

This often-maddening state of affairs triggered an analogy that kept popping into my Excel-weary brain. On a windy summer day near Har-vard Square, I met Andrea Loehr, a blond German who was volunteering as a data cruncher for the PGP and who had been to the South Pole sev-eral times. She was trained as a cosmologist and explained that Antarctica affords astronomers a view that can't be had anywhere else on earth: if one were interested in looking into space, she said, the South Pole was a great place. "It's high, it's dry, and the atmosphere is extremely stable." The problem—and the reason why she decided to switch from stargazing to whole-genome sequence data analysis—is that in cosmology, getting meaningful answers can take decades. "In fact, I may not live to see them," Andrea said. "I like the idea of a project that will have at least some return within a few years."[13]

A few months earlier Hugh Rienhoff had brought up Antarctica to me for a different reason. He had been doing some consulting for Knome and had seen a few of the company's first customers, the ones who had shelled out $350,000 to get their complete genomes done. "What mo-tivated them?" I wondered. "It's curiosity," he said. "People want to go to Antarctica. Why? They like to do it. It's icy, it's cold, it's windy, it's dangerous. People getting their genomes sequenced is that kind of thing. Though I don't think they're learning much they didn't know already."[14]

Or perhaps they were.

When most people think of Henry Louis "Skip" Gates, Jr., chances are they think of his arrest for disorderly conduct in the summer of 2009, the surrounding kerfuffle, and his subsequent beer with the arresting officer and President Obama.[15] They may know that he is the Alphonse Fletcher University Professor and the director of the W. E. B. Du Bois Institute for African and African American Research at Harvard University. That he is a dignified and frequently feted scholar, the winner of a MacArthur Genius Grant.[16] They might also think of his *African American Lives* and *Faces of America* documentary series,[17] which seem to commandeer PBS every other February (certainly more fun than pledge drives), and the famous Americans who participate in them.

What they may not realize is that Skip Gates is hilarious and a keen student of genetics.

As we chatted in the kitchen of his immaculate (and, in the wake of death threats, now quite secure) yellow house, a stone's throw from Harvard's sprawling and semistately campus, Gates made himself a high-fiber, low-fat tuna salad sandwich while I sat at the far end of the large white island in the middle of the room. Soon the doorbell rang. He looked at me.

"If it's the police, tell 'em I ain't here," he deadpanned.[18]

Gates came to see molecular genetics as a tool to uncover African American ancestry in 2000, when he connected with Rick Kittles, then a geneticist at Howard University just beginning molecular studies of African ancestry.[19] But Gates's interest really began at age nine, "the day my grandfather was buried." The next day he began work on the Gates family tree. When Alex Haley's *Roots* was broadcast in 1977, Gates was riveted. As he got older, he became a devoted Africanist, eventually visiting twenty-one countries on the continent. When Kittles explained what he was doing, "I was *down*," said Gates. "I said, 'You're gonna take a little blood and tell me what tribe I'm from?' That's what I'm talkin' about." He soon donated blood.[20]

Time went by and Gates called Kittles. "Hey, man. Where's my Kunta Kinte moment?"[21] In 2000 the database of genetic markers was a shadow of what it is today. And in the relatively early research phase of his studies, Kittles looked only at markers common among Africans. Consequently he could not identify a strong African link to Gates's maternal line.[22]

"I thought it was a simple test: you take some spit and a tribe lights up in Africa," Gates recalled. "I didn't know about private and public genetic databases. It wasn't my field. But some of the time I'm smart enough to know what I don't know. I realized it was time for me to pull back, not treat the subject cavalierly, study the science myself, and surround myself with a battery of experts, some of whom disagreed with Rick."[23]

Subsequent analyses all indicated that Gates's roots were in Europe, and the British Isles in particular. Enthralled with the science of heredity, Gates took the opportunity to school himself. He took the PGP exam and a DVD-based course in undergraduate genetics. He asked the Broad Institute's David Altshuler, Mark Daly, and Eric Lander to mentor him.[24] And George Church, too. "He had an endless series of questions and at times it seemed like it was an infinite loop," recalled George with a smile. "Haploid this and diploid that, 'Were genes like twenty thousand volumes in a library?' and ten other analogies going simultaneously. He seems to have only one setting on his potentiometer, which is *ecstatic*. I think it's going to be a good thing to have him involved."[25]

But involved in what exactly? How did genetic ancestry testing come to be and where is it going?

The consensus—both from fossils and genetic variation studies—is that modern humans originated in Eastern Africa 160,000 to 200,000 years ago.[26] Some of us left Africa for Eurasia, Oceania, and the Americas 60,000 to 100,000 years ago. If you look at the millions of human SNPs that have been characterized, Africa is the most genetically diverse place on the planet. All human genetic variation is a subset of what's in Africa. As a general rule, the farther you get from Africa, the less genetic diversity you see. Genetic differences between human populations are small; however, they are undeniably real. The more markers you look at, the easier it is to distinguish an "Asian" genome from an "African" or "European" one.[27] As scientists identified more markers and typed increasing numbers of individuals from different populations, the prospect of spitting in a tube and learning about one's genetic ancestry became less of a dowsing rod and more of a bona fide science.

Traditionally, genetic ancestry testing was based on Y-chromosome and mitochondrial DNA (mtDNA) markers. The Y chromosome is passed only from father to son. Mitochondrial DNA resides in the mitochondria, the "cellular power plants" that are thought to have evolved from bacteria. Part of this evolutionary artifact is a small circular genome that codes for just a handful of genes. It is passed only from mother to child: the spermatozoa jettison their mitochondria and therefore males do not transmit mitochondrial DNA. Y-chromosome and mtDNA markers are called lineage markers. When sperm meets egg, they do not get shuffled (recombine) the way other chromosomes do because eggs don't have Y chromosomes and sperm don't have mitochondria with which to pair up and exchange parts. Thus they make it easy to reconstruct maternal (mtDNA) and paternal (Y) lineages. They can provide regionally specific information: your paternal line may have roots in Native American populations, let's say. Your maternal line may suggest a Northern European origin. The problem with these markers is they can paint a skewed picture: together mtDNA and the Y chromosome account for less than 1 percent of the human genome. And they are, by definition, linear: they follow a *single* line backward. But of course our family trees only get wider as we move back in time: we have four grandparents, eight great-grandparents, sixteen great-great-grandparents, and so on. At ten generations, we have 1,024 ancestors. A Y-chromosome or mtDNA test will take us directly back to only one of them.[28]

More recently, with the advent of millions of SNPs scattered across the genome and better computer algorithms, genetic genealogists have availed themselves of *autosomal* markers, that is, DNA markers on chromosomes 1 through 22 that travel in pairs and that do get shuffled at every conception. These tests incorporate information from the full range of one's ancestors, not just those in the direct maternal or paternal lines. They suggest, for example, that my recent genetic ancestry is 99 percent European—that is, boring.

In 2009, 23andMe began beta testing of Relative Finder, a program that compared customers' DNA with each other.[29] When two people shared identical segments of DNA, this indicated that they shared a recent common ancestor. The length and number of these identical segments were used to predict the relationship between any two people. As I wrote this, there were fifty other 23andMe users who were predicted to be my third cousins. Did that mean we actually shared great-great-grandparents? Absolutely not. But we were the functional equivalent of third cousins: we shared the same amount of DNA *as if* we had the same great-great-grandparents.

Skip Gates had similar analyses done for himself and his *Faces of America* participants by scientists at the Broad Institute. He was an unabashed fan of this approach. "It's the emotional high point of the series," he said. If you tell somebody, 'You [and this other person] are descended from a common ancestor twenty thousand years ago,' they'll go, 'Oh yeah. Wow. Big deal.' But if you tell them they descended from a common ancestor since the time of Columbus and maybe as recently as two hundred and fifty years ago, that's heavy, even if they don't look alike. They share an *actual human being* in the recent past."[30]

Many of his colleagues (and, I should say, some of mine) in the humanities and social sciences did not share his enthusiasm for genetic ancestry testing, or even constructing family trees. In a blistering takedown in the journal *PMLA*, the University of Virginia's Eric Lott wrote:

> In *Bell Curve* America, genetic conceptions of personhood
> are bound to be dangerous, and they gibe rather nicely with
> rollbacks of the United States' commitments to closing ra-
> cialized gaps in political and material condition.
>
> Nor, finally, is the reactionary conception of history en-

tailed in familial genealogy helpful in this respect. It radically individualizes the past even as it essentializes racial inheritance, turning up people who did or didn't work against great odds to enable their descendants a better future—and in Gates's chosen subjects, a wealthy and famous future at that.[31]

Less frothily, other folks worried that commercial genetic ancestry testing promoted the fallacious idea that "race" is rooted in our DNA.[32]

Gates was, to say the least, unbowed. Yes, he worried about DNA testing being used to justify the creation of a genetic underclass. But he clearly saw it as a potential for empowerment as well. "There's nothing that can disappoint me in any genetic or genealogical analysis of my ancestry. It provides relief, it provides *answers*. Identity commences as a question. Our genealogical identity and our genetic identity are subsets of one large question: *Who am I?* And each bit of data uncovered by your genealogy, your genetic analysis, provides another small answer to this larger conundrum. So there can be no bad or troubling news. Medically yes, these tests could be very troubling, but not in terms of ancestry. I've been black for fifty-nine years, and other than some radical black nationalists, nobody I've met minds having white ancestry. People just want to know who they were. They'd like to know the circumstances. African Americans want to find out *more* about their complex ancestry, not less.

"You can't be a Luddite. You can't stop it. You have to try to understand it. A lot of humanists and social scientists are like this," he said, putting hands over his ears, closing his eyes, and singing loudly. " 'La la la la la la la la la la!!!' I say to them, you people need to read a genetics book! *The geneticists are not making this up.* We're in the era of the recuperation of biology. We have to become scientifically literate so that we can learn how to *intelligently* challenge the potential abuses. And I'm eternally grateful to Rick Kittles for introducing me to genetics and also to its dangers, perils, and limitations. I've learned a new field to some extent." He paused. "I also know I don't have APOE4."[33]

Gates and his father, Henry Louis Gates, Sr., were among the first four to be fully sequenced by Illumina's $48,000 service.[34] They were the first African Americans and the first identifiable father and son to have their complete genomes done. Gates learned that he had six hundred thousand SNPs that had never been seen before.* "Six hundred thousand," he marveled. "I'm the Skip Chip!"[35]

Gates, his father, and his brother also became PGPers, which made them outliers among African Americans. When asked by George to be among the first ten, Rick Kittles declined. "I think the PGP is wonderful," he told me, "but I can't be that ambassador or take that responsibility on behalf of other African Americans. We're still in the enlightenment phase. This sort of information . . . once it's out, it's out. There are downstream effects I'm not sure I can control."[36] Harvard anthropologist Duana Fullwiley, who was studying the PGP, said blacks' reticence about public genomics should not be a surprise. "The real issue is how this information might be used against African Americans. That comes from knowing the history and how certain groups really do get short shrift." She mentioned DNA dragnets and forensic databases. She was sympathetic to Kittles. "When it is in the public realm it can be utilized in all kinds of ways."[37]

So would Skip go public? His ninety-six-year-old father had already said he would. But Gates *Fils* hedged. I thought maybe he was still smarting from the death threats he received and the paparazzi chasing him after his arrest. But it wasn't that. "I have two daughters," he said. "I have all the information about my genome, but they don't. They would be affected by my decision. If one daughter says no, then that's it. It's gotta be unanimous. It can't be undertaken cavalierly. I will probably make it public unless my daughters have strong objections. The advantage of me making mine public is that it would let people study this relationship between father and son. And particularly with the father pushing a hundred."[38]

He called me into the dining room and sat down at his computer. He called up a lavish PowerPoint series that a scientist at Knome had made for

---

* Obviously as more human genomes were sequenced, this number would drop precipitously.

him. Colorful graphics of the two Henry Louis Gateses' genomes came up with their photographs superimposed. And then, shockingly, his mother's picture appeared superimposed upon *her* genome. Given the availability of Skip's full DNA sequence and his father's, it became a trivial exercise to reconstruct half of his mother's genome (she died in 1987). "Seeing this was so moving," he said softly. "The irony was I did it to immortalize my father, and it resurrected my mother."[39]

Having one's genome sequenced, it seemed to me, included both aspects of a trip to the South Pole. First, with some exceptions, a genome contains a lot of information that simply can't be gotten anywhere else. Someone at risk for Huntington's disease could know decades in advance whether he would suffer a slow, downward spiral in middle age or not. An oncologist could learn whether a cancer patient was likely to respond to a particular drug; a neurologist could anticipate whether someone in pain would be helped by codeine. A person with HIV might learn what to expect regarding the course of his illness. An Orthodox Jewish couple could learn whether they were both carriers of mutations that put their offspring at risk for Tay-Sachs disease, the heartbreaking disorder that kills children before age four. An orphan, adoptee, or donor-conceived child could learn about all kinds of hereditary health risks, carrier states, and likely drug responses that, in the absence of information from her biological parents, she could not find from any other source unless she already exhibited the phenotype.

At the other end of the spectrum was curiosity for curiosity's sake. Skip Gates could get a pretty good idea as to his continental ancestry. Others of us were interested in traits, some of them whimsical or at least not likely to have a profound impact on our lives. Some of us might be able to roll our tongues, taste bitter foods differently from others, or drink coffee to excess with no ill effects. We may have blue eyes or brown eyes, curly hair or no hair, freckles or not. We may flush when we drink alcohol, or not. And because high school science classes somehow failed to kill our interest in genetics completely, we would like to know why. Even if genes can give only a woefully incomplete explanation or one that's subject to radical revision, we would simply like to learn about our own human heritable traits—not at a population level, but at a *micro* level, one that pertains directly to us and, to some extent, our family members. This is personal genomics.

The argument that we'd be better off exercising, losing weight, and quitting smoking than learning about our genomes was a nonstarter with me. Indeed, it's a non sequitur. Of course we'd be better off doing those things! And that would be true whether we knew our APOE genotypes or not, and it would probably be true even when we knew what the vast uncharted regions of our genomes meant, whether we were on the cosmological timescale or the nanosecond DNA sequencing one. And anyway, going to the gym and scrutinizing genomes were hardly mutually exclusive. I suppose what bugged me most was the genetic-determinist assumption behind this charge: that if we paid attention to our genes then we were doing so at the expense of everything else. Is this what happened with cholesterol? What about prostate-specific antigen (PSA) and mammography?

Jamie Heywood, founder of PatientsLikeMe.com, whose brother succumbed to ALS at the age of thirty-seven, said, "Genetics today is a very low-resolution picture, but the only way to get started is to get started."[40] For all of the dead ends, the technical hurdles, the possibilities of unpleasant discoveries, and the deep chasms of uncertainty, I am excited. Not by the prospect of digging into my data every day, but by the collective aggregation of genomes and traits and the years of code-breaking we have ahead of us.

I still see my genome as the least of my problems and I believe that that's true for most of us. Certainly, given our tragic history of eugenic behavior, to recoil at the Awesome Power of Genetics is an understandable reflex. But I don't see willful ignorance as much of a solution. I don't like the idea, for example, that GINA may preclude employee wellness programs from collecting family history information because it amounts to private and dangerous genetic information.[41] This was how we were going to enable prevention?

I believe it is possible to demystify the genome while remaining in awe of it. I want to hitch up my thermal underwear, zip up my parka, don my mittens and snowshoes, and go to the coldest, driest, windiest place on earth. And if it turns out that it's not such an extreme experience after all, then so be it—I am convinced that I will still be smarter because of it. And I will try again tomorrow . . . or in ten minutes. Meanwhile I want anyone else who cares or can make use of the contents of my chromosomes in some constructive way to have at it.

After three years, I had done it: While I'd hardly conquered it, I had come and seen my genome, or at least some tiny fraction of it. I was glad to have cracked open my own cells. But of course I ignored most of their contents. Not because they were toxic, but because in the end, this journey of self-exploration had turned out to be more of a speculative intellectual exercise than a life-changing clinical one. I had arrived at the theater early enough to grab a good seat, but the carpenters were still building the sets. Or maybe they were done—maybe the show had started but we were still in the first scene in the first act and there was still no way to know whether the performance would be any good. Chances are some of it would, but the reviews would dribble out in the coming years.

The critics will be difficult to assuage. The genome has been the subject of as much hype as anything science has ever undertaken. "Personalized medicine," "translational medicine," and "genomic medicine" are the relentless catchphrases, the ubiquitous buzzwords, and all too often, the wishful thinking. Indeed I use them myself: on some level I suppose I have been part of the problem. It could be argued that the PGP, too, has had greater success in the media than it has in the laboratory.[42] So far.

By the time you read this, the novelty and the price tag of a human DNA sequence will both have diminished significantly from when I wrote this, much as they've been doing for the last few years. In 2009, Illumina announced it would begin sequencing individuals' whole genomes for $48,000,[43] one-seventh the cost Knome announced it would charge when it launched its service in late 2007.[44] Not long after, Knome itself dropped its price to $68,500.[45] By September 2009, Kevin Shianna was sequencing human genomes for $23,000. By November, Complete Genomics had delivered a genome for $1,726 in reagent costs;[46] in the coming months the company's per-genome costs would shrink even further.[47] In early 2010, Illumina unveiled the HiSeq 2000, a souped-up sequencer that could deliver 200 billion bases of DNA—two complete human genomes at thirty-fold coverage—in a single run at a cost of less than $10,000 per genome.*[48] At Marco Island that year, Jonathan Rothberg, the man who brought the first next-gen sequencer to market, unveiled a semiconductor-based machine costing $50,000 for decoding subsets of the genome for a few hundred bucks. And it was eminently portable: "You can sequence on the back of a donkey if you want to," said Rothberg.[49] Meanwhile, George had begun giving talks around the country: "The $0 Genome."[50]

An affordable price tag was no longer in doubt. What it all meant still was. Maynard Olson, a gangly man with horn-rimmed glasses, one of the genomics pioneers in both yeast and human, told me that we should not expect much of anything from individual genomes before we had sequenced thousands upon thousands of them. Until then, each one might be a modest technical achievement depending upon how fast, how accurately, and at what cost it was decoded. But for its owner, any individual genome would not be much more than a "just so" story.[51]

It was a difficult argument to refute: our genomes may remain "just so" stories for several years. My own genome was no longer an abstraction for me, but neither was it an immediate revelation. My expectation is that the revelations will come, but their exact nature will continue to surprise us. In late 2009, deCODE scientists reported that in some common diseases, which parent the risk allele came from made a difference. In other words, an allele could predispose someone to type 2 diabetes when inher-

---

* In a bid for global genome domination, the Beijing Genomics Institute immediately bought 128 of the new Illumina machines; http://investor.illumina.com/phoenix.zhtml?c=121127&p=irol-newsArticle&ID=1374343&highlight=.

ited from dad. But the same allele actually *protected against* type 2 diabetes when it was inherited from mom.[52] Male and female genomes are chemically modified in different ways. Thus, personal genomics may depend upon having access to one's parents' DNA. One can imagine how this might complicate things in some families.

I suspect that genomes will be like savings bonds or annuities (insert Wall Street joke here): we have invested in them, but, if we are honest with ourselves, in most cases their abiding value to otherwise healthy people has yet to be demonstrated. Those of us who are impatient (I am looking in the mirror) will continue to try to force the issue and extract returns from those investments sooner rather than later.

But even if that takes a long time, I have no intention of boycotting DNA Day (April 25, the day Watson and Crick's paper appeared in 1953—mark your calendar) just because the marketing of personal genomics has outpaced the science. Nor do I have plans to stop going to my daughters' school with salt, liquid soap, and rubbing alcohol and showing the fourth graders how to use these simple things to extract their own DNA. Thinking DNA is cool, even in the face of its often-limited predictive value, hardly makes one a genetic determinist. Yes, we are fixated on genes in part because we find them to be seductive and reductive explanations for human traits and behavior. But we are also fixated upon them because, like the proverbial drunk looking for his keys under the lamppost, we *can* be: sequencing genomes is a helluva lot cheaper and easier than collecting detailed phenotypic data and probably always will be. In the next few years, a genome is likely to be a onetime test one gets at birth—hell, even Francis Collins says so![53] Meanwhile, stress tests, mammograms, and colonoscopies will probably always require repeat visits to the clinic throughout our adult lives.

When my kids ask me about pea plants and fertilization and Watson and Crick and Rosalind Franklin and SNPs, I will tell them what I can. But I will also remind them about *environment*: about the microbiome, about the air they breathe, about diet and exercise, and about how much their mother and I love them.

Here Is a Human Being

Glenn Close has had her genome sequenced. So has Desmond Tutu. So have a few identifiable nuclear families.[1]

And at the beginning of 2010, the Personal Genome Project had 12,500 people in the queue waiting to be participants. Of those, more than a thousand had passed the exam with perfect scores and signed the scary consent form.[2]

All the PGP needed now was money.

Over lunch at the Prudential Mall in Boston between sessions of the first annual Consumer Genetics Conference in 2009, George, Jason, and the PGP's pro bono lawyers from Charlotte, North Carolina, brainstormed about how to keep the project afloat in the absence of much public funding and an unpredictable flow of corporate dollars. George—his usual glass-of-water lunch before him—admitted he worried about it a little bit, though clearly not as much as everyone else at the table did. "It's kind of like the skydiver without a parachute," he said. "Halfway down he says, 'So far so good.' "[3]

In April 2010, there would be a personalgenomes.org thousand-dollar-a-plate fund-raiser/conference featuring many if not most of the world's identified full genomes. Thanks to David Goldstein and Kevin Shianna, I would not have to sneak in through the kitchen. There was some skepticism from the media, but the participants were fully engaged and mostly buoyant. One highlight was John Halamka, clad in a one-of-a-kind black Kevlar dress suit, asking guests to pour water on him only to have it roll right off his synthetic clothes.

Another strategy the PGP adopted was to ask would-be participants for a financial pledge, albeit with the understanding that one's ability to pay would not have any bearing on whether she was accepted as a participant. George and Jason took pains to make sure that participants could not ever construe this arrangement as any sort of a quid pro quo. "A pledge-type arrangement may contribute toward mitigating the coercive lure of a free genome sequence," reasoned Jason. "Recall that genome sequencing had a street value of $350,000 not so long ago. That's a big lure."[4] I understood this argument, but was a $0 genome going to be a lure? Also, I couldn't help but worry that asking for money would create expectations. If I give my local NPR station $100, I expect a coffee mug or a tote bag. If I give it $10,000, then I expect Brooke Gladstone to bring me breakfast in bed, read me the news in a throaty whisper, and pick out my socks.

The heat on the personal genomics companies had been turned down, at least for a moment. In 2009 a paper appeared in Jim Evans's journal *Genetics in Medicine* that included both academics and company officials as coauthors. It laid out a framework for how genetic tests should be evaluated and even paid obeisance to the idea of "personal utility" as put forth by Robert Green.[5] At first glance this publication seemed like a "Kumbaya moment." If nothing else, it was a clarion call for more research monies and "independent panels" to study the positive and negative effects of consumer genomics. At the end of the paper was a list of the commercial coauthors' conflicts of interest. I found it curious, though not terribly surprising, that while none of the academic physicians and scientists who stood to benefit from an influx of research dollars had any qualms about citing their private-sector colleagues' conflicts of interest, they never felt conflicted enough to include their own names on the list.

At least two of the seminal figures in commercial personal genomics would be witnessing this new era of public-private cooperation from different vantage points. Linda Avey left 23andMe in 2009 to start a nonprofit foundation devoted to research in Alzheimer's disease, which claimed her father-in-law.[6] It was not clear to me whether she jumped or was pushed, though I have my suspicions. And Navigenics cofounder Dietrich Stephan opted to step away from an operational role in the company. Once Navigenics moved out of the "invention" stage and became a more conventional business trying to move spit kits, Dietrich was ready to return to academia. Indeed, as early as the SoHo launch party, he seemed to me to be tired, beaten down by his company's legions of critics: medical geneticists, regulators, ethicists, genetic counselors. "I went from being a respected academic researcher who got grants like nobody's business," he said, "to landing in a world that was totally foreign to me and having my scientific credibility called into question. It was like getting kicked in the stomach every day."[7]

"There's something about this field," agreed Linda. "People love to shoot arrows at us."[8]

Dietrich started a research institute in the D.C. area, the Ignite Institute,[9] which immediately jumped into the fray and ordered a hundred of the latest SOLiD sequencers.[10] He had no regrets about the arrows or the blows to the stomach. "I think history will look back kindly on us: Navigenics launched direct-to-consumer genomics for chronic disease in a gold-standard way. deCODE has done a great job on the science. . . . I just wish they were nicer. Kari's sharp elbows are unnecessary. I think 23andMe's greatest skill has been in listening and iterating: they find out what regulators and customers want, and they go back home and change things."[11]

As I put this book to bed, it was still not clear what the fate of the Big Three would be. With the meltdown of the Icelandic economy and the company's failure to develop any genome-based blockbusters, deCODE was in real trouble. In November 2009 it filed for bankruptcy, although the ever-pugnacious Kari Stefansson vowed to fight on.[12] In early 2010, the company emerged as a "consulting" firm whose specialty was genome interpretation, while the fate of its personal genome scanning business remained uncertain.[13] Meanwhile, 23andMe announced a series of layoffs.[14] And the specter of harsh regulation was never far away.

Indeed, this was never clearer than when newcomer Pathway Genomics announced it would begin selling DNA spit kits at Walgreens drugstores in May 2010. The bioethics community had its usual conniption, and within a few days the FDA put the kibosh on the whole thing.[15] Shortly thereafter Congress sent letters to Pathway, Navigenics, and 23andMe requesting all kinds of information.[16]

But neither was the news all bad. Right before Christmas, 23andMe announced that it had bagged another $14 million, bringing its 2009 total to nearly $28 million.[17] In early 2010, I sat on a panel with Kari Stefansson and Dietrich Stephan. Kari had his new business partner in tow and his elbows seemed to me to be less sharp. Dietrich seemed relieved to be free of the day-to-day headaches of running Navigenics, though he joked that in his new role running an institute he had exchanged doing battle with regulators and academics for begging for money from hospital administrators.

In my view, the personal genomics companies' success would ultimately hinge on doctor buy-in, more predictive science, and something that went beyond specific predictions: a cumulative sense that this information was useful in making decisions about one's life and health. They would also need a deep well of creative and tireless people who have problems with authority; people like Dietrich Stephan, Kari Stefansson, and Linda Avey.

I was pretty sure that I would continue to pester PGP staffers about my latest data releases. Not to be a jerk, but because I actually wanted to have a relationship with them. I wanted them to know that there was *someone*—an anxious middle-aged dude with a wife and two kids, susceptible to metabolic syndrome, plays a crude approximation of guitar in a rock band— behind those cells, that tissue, those brain scans, and that DNA, and that he expected the PGP to live by its own self-imposed code. *I cared.* Serving on the IRB had made me aware of how often investigators lost sight of whom they were studying and why. I had—and have—every reason to think the PGP would be different.

I had swallowed open consent whole. I wondered if George might yet live to regret it . . . if he ever regretted anything.

"I'm still at the 'pre-gret' stage," he told me in an email with a subject line that read "RE: gret." "I suppose that dissecting the past can help with planning the future . . . so my regrets are for opportunities missed in the future despite warnings in the present. There are a vast number of alternative universes: perhaps in some my career goes better, but global warming goes worse."[18]

As I sat in the audience in a chilly lecture hall in Boston and listened to him rhapsodize yet again about the $0 genome (and all of the other cheap "omes" headed our way), I was struck anew by his tirelessness and optimism. Only George could have had the audacity and the naïveté, the cojones and the guilelessness, to write an article in *Scientific American* in January 2006 called "Genomes for All."[19] It was a declaration that while we may have only the vaguest of clues as to what the genome means or what it does, like Roald Amundsen headed south, we were ready to find out for ourselves. Not because we have any illusions about genes as the secrets of our being.

But because we don't.

Acknowledgments

The book you hold in your hands or on your screen could not have happened without the extraordinary generosity of countless loved ones, friends, colleagues, and strangers. I will try to count them anyway.

I am eternally grateful to George Church for allowing me to shadow him for four years and never once telling me to go away, even though I gave him ample justification to do so on many occasions.

I thank Hunt Willard for his seemingly limitless supply of faith and patience. I don't know many people who could say that about their bosses. I have been blessed for many years in this regard.

I am deeply indebted to David Goldstein, not only for agreeing to sequence me, but for supporting me in so many other tangible and intangible ways.

Thank you to my agent, Heather Schroder, for her competence, unflinching support, and tough love. Thanks also to Nicole Tourtelot for able assistance.

I am terribly grateful to my first editor, Elisabeth Dyssegaard, and to her successor, Bill Strachan, at HarperCollins. Thanks also to the aptly named Kate Whitenight (though I would have spelled it with a *k*).

My colleague and friend, Bob Cook-Deegan, has been an invaluable source of advice, support, insight, and wisdom. I owe him much.

To Duncan Murrell, who has taught me about writing, about editing, and about humanity: thanks, D.

Everyone attached to the Personal Genome Project was extremely helpful and I owe them more than they'll ever know, especially Jason Bobe, Jeantine Lunshof, Dan Vorhaus, Joe Thakuria, Margret Hoehe, Robert Green, Marilyn Ness, and Dana Waring Bateman. I am extremely grateful to Terry Bard, Abraham Rosenbaum, Sasha Wait Zaranek, Kevin McCarthy, and Andrea Loehr. Other Church lab members to whom I owe thanks: Madeleine Ball, Tom Clegg, Jin Billy Li, Jay Lee, John Aach, Mike Chao, Rich Terry, Mayra Mollinedo, and Mark Umbarger. And also to former Church lab members who shared a lot with me: Jay Shendure, Greg Porreca, Yuan Gao, Kun Zhang, Jun Zhu, and Xiaodi Wu. Many thanks also to Harvard's Isaac Kohane and Alexandra Shields. Thanks also to Goodwin Procter LLP.

And to the two most important members of Team Church—Ting Wu and Marie Wu—I say thanks for your generosity.

Without exception, the PGP-10 were both forthcoming and articulate and I feel fortunate to have gotten to hang around with them: John Halamka, Esther Dyson, Kirk Maxey, Steven Pinker, James Sherley, Rosalynn Gill, Keith Batchelder, and Stan Lapidus.

Stan also spoke to me at length about Helicos BioSciences, as did many other past and present employees in his company. Steve Lombardi always made time for me and for that I thank him profusely. Thanks also to Tim Harris for his time and perspective. Bill Efcavitch, John Boyce, Patrice Milos, and Sepehr Kiani were both gracious and helpful.

Walter Gilbert shared his intimate knowledge of science, art, and George Church. He is also responsible for the title of this book. Thank you, Dr. Gilbert! And thanks to Deb Munroe.

I owe many thanks to Francis Collins and to the true locus of power at NIH, Kathy Hudson ☺. At NHGRI, I thank Eric Green, Les Biesecker, and Jeff Schloss. Thanks also to Sharon Terry for her thoughts and for her heroic work with the Genetic Alliance.

In Boston, a huge thank-you goes to Kay Aull. Thanks to George Annas and Winnie Roche. Thanks very much to Jorge Conde and Julie Yoo at Knome. Thanks also to Heidi Rehm at the Partners HealthCare Center for Personalized Genetic Medicine and to Jamie Heywood and PatientsLikeMe.

I appreciate all of the time and effort expended on my behalf by Mike Cariaso at SNPedia and the time spent with his partner in crime, Greg Lennon.

Thank you to everyone in the Bay Area who made time and space for me. Thank you to the Rienhoff-Hane family for its deep well of generosity: Hugh, Lisa, Colston, Mac, and Beatrice. Thank you to Hank Greely for his time and insight. Heartfelt thanks to Dietrich Stephan, Sean George, Elissa Levin, Michael Nierenberg, Eran Halperin, Michelle Cargill, Steve Moore, Vance Vanier, and Mari Baker. I am more than grateful to Ryan Phelan, Lisa Lee, Lisa Kessler, Anne Wojcicki, Joanna Mountain, Matt Crenson, Andro Hsu, and Darren Platt. Thanks also to Mildred Cho and Sandra Soo-Jin Lee. Thanks for dinner and more, Steve Brenner. Thank you to Un Kwon-Casado, Eli Casdin, and, in Boston, John Sullivan. Thanks also to Jeff Gulcher and Kari Stefansson. Special thanks to Linda Avey, Jenny Reardon, and Barbara Prainsack.

To my colleagues at Duke, I am deeply grateful. These people include Kevin Shianna and Dongliang Ge, both of whom were indispensable. I am indebted to Kendall Morgan, Kristen Linney, Anna Need, Jason Smith, Linda Hong, Ryan Campbell, Jessica Maia, Abanish Singh, Tom Urban, Mingfu Zhu, Jacques Fellay, Liz Cirulli, Kim Pelak, Elizabeth Ruzzo, Curtis Gumbs, Erin Heinzen, Angie Cherry, Charmaine Royal, Chris Heaney, Subashini Chandrasekran, Sara Katsanis, Christine Oien, Alan Cowles, Sabrina Australie, Georgia Barnes, Jay Hamilton, Shelley Stonecipher, Ken Rogerson, Julianne O'Daniel, John Harrelson, Sharon Ellison, John Falletta, Geoff Ginsburg, Rob Mitchell, Amy Laura Hall,

Priscilla Wald, Greg Wray, Philip Benfey, Dan McShea, Alex Rosenberg, Alex Cho, Christina Kapustij, Susan Brooks, Lauren Dame, Susanne Haga, Laura Beskow, Tomalei Vess, Lynne Skinner, Chris Tobias, Amy Fowler, Tom Burke, Cathy Sciambi, Donna Crutchfield, Ellen Brearley, Cindy Wicker, Simon Gregory, and my dear friend/philosopher/PERL maven, Mark DeLong.

And to my collaborators from elsewhere: Louiqa Raschid, Ritu Agarwal, Samir Khuller, and Brad Malin—thank you for including me.

Warm thanks to Maynard Olson, Oliver Smithies, Sam Levy, Vera Rubin, Craig Venter, and Heather Kowalski.

To everyone involved with Project Jim, thank you very much: Jim Watson, Jonathan Rothberg, David Wheeler, Michael Egholm, Richard Gibbs, and especially Amy McGuire.

Thank you, Jim Evans and Gail Henderson—see you at Glasshalfull. Much gratitude to Hal Dietz, a gifted physician and researcher, and a mensch. Thank you to Tiffany Marum for her lack of cynicism. Thanks to Bob Davis and Peter Whitehouse. Special thanks to Mary-Claire King.

Thank you to Skip Gates, aka The Man, for his generosity and time. Thank you to his colleague at Harvard and my friend Duana Fullwiley. Many thanks to Rick Kittles. Thanks also to Amy Gosdanian and Abby Wolf.

Chad Nusbaum was incredibly helpful; I hope I have not damaged his career. Others at the Broad Institute who were kind enough to speak with me include Stacey "Noodles" Gabriel, Pardes Sabeti, and Carsten Russ.

The sequencing and genome biology community was rife with fascinating and generous people. These included Rade Drmanac, Jennifer Turcotte, Andy McCallion, Elaine Mardis, Steve Turner, Trevin Rard, Kevin McKernan, Alan Blanchard, Steve McPhail, Jay Flatley, Tristan Orpin, Ian Goodhead, Steve Quake, Evan Eichler, Hugues Roest-Crollius, Anne Pontius, David Bentley, Mostafa Ronaghi, Eddy Rubin, Richard Fair, Pauline Ng, Zhen Lin, Nelson Axelrod, and Zhuo Li.

Many thanks go to my graduate adviser and friend Aravinda Chakravarti for the years of wisdom, openness, and hospitality.

Thank you, friends: Cathy Olofson, Tara Matise, Sarah Shaw Murray, and Sue Slaugenhaupt.

Thanks to John Inglis at Cold Spring Harbor. Thank you to Wendy Kramer and the Donor Sibling Registry. Thanks to Bill Catalona for his time and insight and to Cissy Lacks. Thank you also to Deb McDermott.

Thank you to the indefatigable staff at *Genome Technology*, especially Julia Karow and Meredith Salisbury. Thank you also to Kevin Davies of *Bio-IT World* and Orli Bahcall and Myles Axton at *Nature Genetics*.

I am indebted to the genome bloggerati, especially Daniel Macarthur, Blaine Bettinger, Hsien Hsien Li, Jonathan Eisen, and Steve Murphy. Thanks also to Bora Zivkovic, David Kroll, and Sheril Kirshenbaum.

Many thanks to the Genetics & Public Policy Center and to Paul Easton.

Thanks to my writer friends, whose support never flagged: Rich Remsberg, Barry Yeoman, Carl Zimmer, Rebecca Skloot, Richard Ziglar, David Dobbs, David Ewing Duncan, Amy Harmon, and Thomas Goetz. Thanks to Nicole Chaison and Jen Bergmark. I offer my humble gratitude to Theresa Rebeck for all kinds of stuff.

Warm thanks to Sea Cow, the best band a guy could have: DJ, Jennie, Shoney, and Sneezy. Thanks also to Durham's answer to Martha Stewart, Julie Maxwell.

Finally, I could not have written this book without the love, help, and buy-in of my family. Josh and Mira Angrist and their kids Noam and Adie housed and fed me time and again and tolerated my chronic messiness. Ezra Angrist and Michele Penner gave of themselves, their children Jordan and Jesse, and their stories. My parents, Sarah and Stan, in addition to sharing their genomes with me and with my brothers, told me about our family history and never wavered in their support. Bill Rebeck put me up and put up with me for many nights in Washington, D.C. I will have to find other excuses to visit and annoy him.

And to Ann and to the best things ever connected to my genome, Stella and Lena: thank you, girls. Again.

Notes

A WORD ABOUT VERACITY

This book is true, it is nonfiction, which is to say that I didn't make anything up or change any names. I made a good-faith effort to honestly and accurately portray people and events as I experienced them. Interviews were recorded digitally or are derived from detailed notes during and/or after conversations and meetings. I edited dialogue for clarity or grammar and in a few instances I spliced together statements from the same person that were made at different times. In rare cases (e.g., the James Sherley affair), I relied more heavily on secondary sources, be they journalistic or scholarly. Wherever possible, sources are end-noted and all web links were active at the time this book was being edited. Where first-person speculative statements and opinions are offered without attribution or quotation marks, they are mine and mine alone. Please don't hassle the nice people at Duke University or HarperCollins on account of my *mishegas*.

EPIGRAPH

1. D. J. Kevles and L. E. Hood, *The Code of Codes: Scientific and Social Issues in the Human Genome Project* (Cambridge, Mass.: Harvard University Press, 1992).
2. J. W. v. Goethe and D. J. Enright, *The Sayings of Goethe* (London: Duckworth, 1996).

3. Seen in George Plimpton's diary by James Scott Linville: http://themain point.blogspot.com/2009/02/more-on-plimpton.html.

CHAPTER 1: THE NUMERATOR

1. http://www.myspace.com/seacowrocks.
2. personalgenomes.org.
3. P. Dizikes, "Your DNA is a snitch," Salon.com, February 17, 2009.
4. P. Szekely, "Railroad to pay $2.2 million in DNA test case illegally testing workers for genetic defects," Reuters, May 8, 2002.
5. *Norman-Bloodsaw v. Lawrence Berkeley Laboratory,* Fed Report, 1998, 135, pp. 1260–76; S. French, "Genetic testing in the workplace: The employer's coin toss," *Duke Law & Technology Review,* September 5, 2002, p. E1.
6. S. Greenhouse and M. Barbaro, "Wal-Mart memo suggests ways to cut employee benefit costs," *New York Times,* October 26, 2005.
7. K. Kaplan, "U.S. military practices genetic discrimination in denying benefits," *Los Angeles Times,* August 18, 2007.
8. S. Baruch and K. Hudson, "Civilian and military genetics: Nondiscrimination policy in a post-GINA world," *American Journal of Human Genetics,* 2008, 83(4): 435–44.
9. K. L. Hudson, M. K. Holohan, and F. S. Collins, "Keeping pace with the times—the genetic information nondiscrimination act of 2008," *New England Journal of Medicine,* 2008, 358(25): 2661–63.
10. L. A. Cupples, L. A. Farrer, A. D. Sadovnick, N. Relkin, P. Whitehouse, and R. C. Green, "Estimating risk curves for first-degree relatives of patients with Alzheimer's disease: The REVEAL study," *Genetics in Medicine,* 2004, 6(4): 192–96.
11. A. L. McGuire and M. A. Majumder, "Two cheers for GINA?" *Genome Medicine,* 2009, 1(1): 6; M. A. Rothstein, "Putting the Genetic Information Nondiscrimination Act in context," *Genetics in Medicine,* 2008, 10(9): 655–56.
12. S. Reitz, "Connecticut woman alleges genetic discrimination at work," AP, April 1, 2010.
13. R. N. Haricharan and K. E. Georgeson, "Hirschsprung disease," *Seminars in Pediatric Surgery,* 2008, 17(4): 266–75.
14. A. J. Burns and V. Pachnis, "Development of the enteric nervous system: Bringing together cells, signals and genes," *Neurogastroenterology & Motility,* 2009, 21(2): 100–2.

15. F. de Lorijn, G. E. Boeckxstaens, and M. A. Benninga, "Symptomatology, pathophysiology, diagnostic work-up, and treatment of Hirschsprung disease in infancy and childhood," *Current Gastroenterology Reports*, 2007, 9(3): 245–53.

16. J. Kessmann, "Hirschsprung's disease: Diagnosis and management," *American Family Physician*, 2006, 74(8): 1319–22.

17. Localizing major disease susceptibility genes to a chromosomal neighborhood was still a big deal in the early 1990s. Today, thanks in large part to the Human Genome Project, most of that work has been done. If we want to learn about a gene involved in disease, we usually just have to turn on our computers and read. M. Angrist, E. Kauffman, S. A. Slaugenhaupt, et al., "A gene for Hirschsprung disease (megacolon) in the pericentromeric region of human chromosome 10," *Nature Genetics*, 1993, 4(4): 351–56; S. Lyonnet, A. Bolino, A. Pelet, et al., "A gene for Hirschsprung disease maps to the proximal long arm of chromosome 10," *Nature Genetics*, 1993, 4(4): 346–50.

18. J. Amiel, E. Sproat-Emison, M. Garcia-Barcelo, et al., "Hirschsprung disease, associated syndromes and genetics: A review," *Journal of Medical Genetics*, 2008, 45(1): 1–14.

19. M. Angrist, S. Bolk, B. Thiel, et al., "Mutation analysis of the ret receptor tyrosine kinase in Hirschsprung disease," *Human Molecular Genetics*, 1995, 4(5): 821–30.

CHAPTER 2: "HIS OWN DRUM"

1. Interview with John Halamka, July 18, 2007.

2. Interview with John Halamka, October 26, 2006.

3. http://geekdoctor.blogspot.com/2008/01/engineering-eye-for-average-guy_16.html.

4. Interview with John Halamka, July 18, 2007.

5. Interview with John Halamka, October 26, 2006.

6. Ibid.

7. From the Personal Genome Project Consent Form, 10th Draft, July 27, 2006.

8. C. R. Scriver, "After the genome—the phenome?" *Journal of Inherited Metabolic Disease*, 2004, 27(3): 305–17.

9. http://www.bccresearch.com/report/BIO045A.html.

10. E. Pettersson, J. Lundeberg, and A. Ahmadian, "Generations of sequencing technologies," *Genomics*, 2009, 93(2): 105–11.

11. Interviews with George Church, February 9 and March 13, 2007.

12. http://www.personalgenomes.org/public/1.html.

13. Interview with George Church, February 2, 2009.

14. http://nobelprize.org/nobel_prizes/medicine/laureates/2007/index.html.

15. http://www.genome.duke.edu/press/genomelife/archives/issue09/GL_Aug04.pdf.

16. Interview with George Church, August 11, 2008.

17. J. Kirshenbaum, "Top hat, white tie and bare toes," *Sports Illustrated*, January 4, 1971.

18. http://www.waterskihalloffame.com/docs/Bios/Stew%20McDonald.htm.

19. Interview with George Church, October 20, 2008.

20. Interview with Ting Wu, July 17, 2007.

21. Interview with George Church, October 25, 2006.

22. G. M. Church, J. L. Sussman, and S. H. Kim, "Secondary structural complementarity between DNA and proteins," *Proceedings of the National Academy of Sciences,* 1977, 74(4): 1458–62.

23. Interviews with Walter Gilbert, January 16 and 23, 2007.

24. A. M. Maxam and W. Gilbert, "A new method for sequencing DNA," *Proceedings of the National Academy of Sciences*, 1977, 74(2): 560–64.

25. G. B. Kolata, "The 1980 Nobel Prize in chemistry," *Science,* 1980, 210(4472): 887–89.

26. Interview with Ting Wu, July 17, 2007.

27. Ibid.

28. Interview with George Church, October 25, 2006.

29. http://thepersonalgenome.com/2007/08/nom-de-ome-a-pseudonym-for-your-genome/.

30. Interview with George Church, October 25, 2006.

31. D. Kennedy, "Not wicked, perhaps, but tacky," *Science*, 2002, 297(5585): 1237.

32. S. Levy, G. Sutton, P. C. Ng, et al., "The diploid genome sequence of an individual human," *PLoS Biology*, 2007, 5(10): e254.

33. http://www.genomeweb.com/2003-genome-technology-all-stars.

34. R. D. Mitra, J. Shendure, J. Olejnik, O. Edyta Krzymanska, and G. M. Church, "Fluorescent in situ sequencing on polymerase colonies," *Analytical Biochemistry*, 2003, 320(1): 55–65.

35. http://www.genomeweb.com/sequencing/danaher-ship-first-church-lab-polonator-week-presenting-instrument-agbt.

36. Interviews with George Church, February 9 and March 13, 2007.

37. Interview with George Church, October 26, 2006.

38. http://arep.med.harvard.edu/PGP/Anon.htm.

39. http://ncvhs.hhs.gov/980128tr.htm.

40. A. Motluk, "Anonymous sperm donor traced on Internet," *New Scientist*, November 3, 2005.

41. Z. Lin, R. B. Altman, and A. B. Owen, "Confidentiality in genome research," *Science*, 2006, 313(5786): 441–42.

42. N. Homer, S. Szelinger, M. Redman, et al., "Resolving individuals contributing trace amounts of DNA to highly complex mixtures using high-density SNP genotyping microarrays," *PLoS Genetics*, 2008, 4(8): e1000167.

43. http://grants.nih.gov/grants/gwas/data_sharing_policy_modifications _20080828.pdf.

44. Interview with George Church, June 6, 2007.

45. Interview with George Church, October 27, 2006.

46. http://arep.med.harvard.edu/gmc/diet.html.

47. Interview with George Church, August 11, 2008.

48. Interview with Isaac Kohane, October 26, 2006.

49. http://www.1000genomes.org/page.php.

50. Interview with George Church, October 26, 2006.

51. A. Regalado, "Entrepreneur puts himself up for study in genetic 'tell-all,'" *Wall Street Journal*, October 18, 2006.

52. http://jimwatsonsequence.cshl.edu/cgi-perl/gbrowse/jwsequence/.

53. P. Sherwell, "DNA father James Watson's 'holy grail' request," Telegraph .co.uk, May 10, 2009.

54. Interview with James Watson, May 11, 2007.

55. H. Nugent, "Black people 'less intelligent,' scientist claims," *Times of London* online, October 17, 2007.

56. J. Crace, "Double helix trouble," Guardian.co.uk, October 16, 2007.

57. Interview with George Church, February 9, 2007.

58. Interview with Terry Bard, November 8, 2006.

59. Ibid.

60. Ibid.

61. Anonymous, February 22, 2007.

62. Email correspondence between Jeff Schloss and George Church, May 6, 2006.

63. T. A. Manolio and F. S. Collins, "The HapMap and genome-wide association studies in diagnosis and therapy," *Annual Review of Medicine*, 2009, 60: 443–56.

64. http://www.hapmap.org/downloads/elsi/consent/Consent_Form_Template. doc.

65. Personal Genome Project Informed Consent Form, 10th draft, July 27, 2006.

66. S. Olson, "Who's your daddy?" *Atlantic*, July/August 2007.

67. G. J. Annas, L. H. Glantz, and P. A. Roche, "Drafting the genetic privacy act: Science, policy, and practical considerations," *Journal of Law, Medicine & Ethics*, 1995, 23(4): 360–66.

68. Interview with Winnie Roche and George Annas, October 26, 2006.

69. R. C. Green and G. J. Annas, "The genetic privacy of presidential candidates," *New England Journal of Medicine*, 2008, 359(21): 2192–93.

70. http://arep.med.harvard.edu/gmc/.

71. http://arep.med.harvard.edu/gmc/pers.html.

72. Interview with Ting Wu, July 17, 2007.

73. Interview with George Church, October 25, 2006.

74. Ibid.

75. http://www.dw-world.de/dw/article/0,,4201588,00.html.

76. Interview with Margret Hoehe, October 10 and 11, 2008.

CHAPTER 3: "WHY SHOULD WE MAKE THEM GO OUT ON THE DANCE FLOOR?"

1. E. Phillips and S. Mallal, "Successful translation of pharmacogenetics into the clinic: The Abacavir example," *Molecular Diagnosis & Therapy*, 2009, 13(1): 1–9.

2. J. Lazarou, B. H. Pomeranz, and P. N. Corey, "Incidence of adverse drug reactions in hospitalized patients: A meta-analysis of prospective studies," *JAMA*, 1998, 279(15): 1200–5.

3. C. Elliott, "Guinea-pigging: Healthy human subjects for drug-safety trials are in demand. But is it a living?" *New Yorker*, January 7, 2008, p. 36.

4. S. D. Warren and L. D. Brandeis, "The right to privacy," *Harvard Law Review*, 1890, 4(5).

5. R. Rao, "Genes and spleens: Property, contract, or privacy rights in the human body?" *Journal of Law, Medicine & Ethics*, 2007, 35(3): 371–82.

6. Interview with William Catalona, June 9, 2008.

7. www.circare.org/lex/03cv01065_pltfposthearingreply.pdf.

8. J. R. Minkel, "Uninformed consent," *Scientific American*, 2006, 295(4): 22, 24.

9. Interview with William Catalona, June 9, 2008.

10. http://vlex.com/vid/catalona-mcgurk-wilard-parron-missios-37825718.

11. T. Herrmann, "8th circuit rules Washington University has ownership of medical research," *Kansas City (Missouri) Daily Record*, June 21, 2007.

12. J. Ritter, "Who owns your body," *Chicago Sun-Times*, September 17, 2007.

13. Interview with William Catalona, June 9, 2008.

14. Ibid.

15. Email from William Catalona, October 22, 2009.

16. Interview with Hank Greely, December 6, 2007.

17. Interview with Amy McGuire, June 15, 2007.

18. Interview with Hank Greely, December 6, 2007.

19. http://grants.nih.gov/grants/gwas/town_hall_mtg/index.htm.

20. H. Skaletsky, T. Kuroda-Kawaguchi, P. J. Minx, et al., "The male-specific region of the human Y chromosome is a mosaic of discrete sequence classes," *Nature*, 2003, 423(6942): 825–37.

21. T. Francis, "Medical dilemma: Spread of records stirs patient fears of privacy erosion," *Wall Street Journal*, December 26, 2006.

22. Email from George Church, January 4, 2007.

23. Interview with George Church, February 9, 2007.

24. Email from George Church, March 1, 2007.

25. E. Check, "Celebrity genomes alarm researchers," *Nature*, 2007, 447(7143): 358–59.

26. D. M. Pressel, "Nuremberg and Tuskegee: Lessons for contemporary American medicine," *Journal of the National Medical Association*, 2003, 95(12): 1216–25.

27. D. Micklos and E. Carlson, "Engineering American society: The lesson of eugenics," *Nature Reviews Genetics*, 2000, 1(2): 153–58.

28. C. Hunt-Grubbe, "The elementary DNA of Dr Watson," *Sunday Times* (of London), October 14, 2007.

29. J. D. Watson, *The Double Helix: A Personal Account of the Discovery of the Structure of DNA* (London: Weidenfeld & Nicolson, 1968).

30. http://nobelprize.org/nobel_prizes/medicine/laureates/1962/.

31. J. Kim, J. M. Basak, and D. M. Holtzman, "The role of apolipoprotein E in Alzheimer's disease," *Neuron*, 2009, 63(3): 287–303.

32. J. D. Watson, *The Double Helix: A Personal Account of the Discovery of the Structure of DNA*, 1st ed. (New York: Atheneum, 1968).

33. "An intellectual entente," *Harvard Magazine*, September 10, 2009.

34. S. Connor, "James Watson, Nobel Prize winner: Welcome to the Watson wonderland," *Independent* (U.K.), February 3, 2003.

35. J. D. Watson, *Recombinant DNA: Genes and Genomes: A Short Course*, 3rd

ed. (New York: W. H. Freeman and Cold Spring Harbor Laboratory Press, 2007).

36. D. A. Wheeler, M. Srinivasan, M. Egholm, et al., "The complete genome of an individual by massively parallel DNA sequencing," *Nature*, 2008, 452(7189): 872–76.

37. E. Singer, "The $2 million genome," *Technology Review*, June 1, 2007.

38. K. Davies, "Project Jim: Watson's genome goes public," *Bio-IT World*, June 13, 2007.

39. Interview with Michael Egholm, May 9, 2007.

40. Interview with Amy McGuire, May 9, 2007.

41. http://biologos.org/news-events/francis-collins-talks-biologos-with-christianity-today/.

42. http://biologos.org/about.

43. http://meetings.cshl.edu/ . . . /2007%20Past%20Programs/Genome%20Program.pdf.

44. N. Mekel-Bobrov, S. L. Gilbert, P. D. Evans, et al., "Ongoing adaptive evolution of ASPM, a brain size determinant in Homo sapiens," *Science*, 2005, 309(5741): 1720–22.

45. Interview with James Watson, May 11, 2007.

46. Interview with George Church, February 9, 2007.

47. Interview with James Watson, May 11, 2007.

48. Ibid.

49. Remarks by James Watson to Biology of Genomes Meeting, Cold Spring Harbor Laboratory, May 11, 2007.

50. Interview with Ting Wu, July 17, 2007.

51. Interview with George Church, April 2, 2007.

52. http://thepersonalgenome.com/category/genetics-profession/.

53. Interview with George Church, June 6, 2007.

54. S. M. Wolf, F. P. Lawrenz, C. A. Nelson, et al., "Managing incidental findings in human subjects research: Analysis and recommendations," *Journal of Law, Medicine & Ethics*, 2008, 36(2): 219–48, 211.

55. J. Murphy, J. Scott, D. Kaufman, G. Geller, L. LeRoy, and K. Hudson, "Public expectations for return of results from large-cohort genetic research," *American Journal of Bioethics*, 2008, 8(11): 36–43.

56. Interview with George Church, October 25, 2006.

57. Ibid.

58. Interview with George Church, February 9, 2007.

59. Interview with Marie Wu, July 17, 2007

60. http://www.thepersonalgenome.com.

61. www.genome.duke.edu/press/genomelife/ . . . /GL_MarApr07.pdf.

CHAPTER 4: "BUT WHEN IS IT MATURE?"

1. Interview with Matt Crenson, December 5, 2007

2. http://spittoon.23andme.com/2008/02/22/23andmore-paternal-ancestry-free-demo-accounts-and-an-expanded-gene-journal/.

3. https://www.23andme.com/you/health/.

4. http://www.techcrunch.com/2008/01/22/1000-free-23andme-kits-for-davos-attendees-plus-one-for-techcrunch-readers/.

5. https://www.23andme.com/gen101/variation/buffett/.

6. http://spittoon.23andme.com/2008/06/16/haplogroups-of-the-rich-and-famous/; http://www.flickr.com/photos/edyson/2228962004/

7. https://www.23andme.com/pregnancy/.

8. G. Y. Tanaka, *Digital Deflation: The Productivity Revolution and How It Will Ignite the Economy* (New York: McGraw-Hill, 2004).

9. R. Osborne, "Perlegen signs AstraZeneca, plus two deals with Pfizer," *Bio-World Today*, January 12, 2004.

10. B. M. Neale and S. Purcell, "The positives, protocols, and perils of genome-wide association," *American Journal of Medical Genetics: Part B: Neuropsychiatric Genetics*, 2008, 147B(7): 1288–94.

11. Interview with Linda Avey, December 5, 2007.

12. Ibid.

13. Ibid.

14. Ibid.

15. http://www.wired.com/wiredscience/2009/03/science-as-seer/.

16. Interview with Linda Avey, December 5, 2007.

17. K. Pattison, "Spit party," *Fast Company,* April 7, 2008.

18. Interview with Dietrich Stephan, December 4, 2007.

19. Ibid.

20. Ibid.

21. Ibid.

22. https://www.navigenics.com/visitor/about_us/press/releases/company_launch_110607/.

23. Interview with Dietrich Stephan, December 4, 2007.

24. S. R. Browning, "Missing data imputation and haplotype phase inference for genome-wide association studies," *Human Genetics*, 2008, 124(5): 439–50.

25. Interview with Elissa Levin, December 4, 2007.

26. http://www.decodeme.com/faq.

27. https://www.23andme.com/you/faqwin/professionals/.

28. http://www.dnadirect.com/web/labs/genetic-counseling-services.

29. K. Davies, "Pathway Genomics joins the direct-to-consumer genomics parade," *Bio-IT World*, July 15, 2009.

30. K. Davies, "Al Gore helps Navigenics launch personal genomics service," *Bio-IT World,* April 9, 2008.

31. P. McGeehan, "On the retail frontier, another shop in SoHo for the person who has everything," *New York Times*, April 13, 2008.

32. Interview with Tara Matise, April 15, 2008.

33. National Society of Genetic Counselors Professional Status Survey, 2008, http://www.nsgc.org/career/pss_index.cfm.

34. Interview with Deb McDermott, April 15, 2008.

35. Interview with Dietrich Stephan, April 16, 2008.

36. "Ready or not," *Nature*, 2008, 452: 666.

37. Interview with Dietrich Stephan, April 16, 2008.

38. R. Langreth and M. Herper, "States crack down on online gene tests," *Forbes*, April 18, 2008.

39. Interview with Dietrich Stephan, April 16, 2008.

40. T. Ray, "Will other states follow NY, Calif., in taking on DTC genetic-testing firms?" *Pharmacogenomics Reporter*, June 25, 2008.

41. R. Langreth, "California orders stop to gene testing," *Forbes*, June 14, 2008.

42. http://www.hhs.gov/news/press/2002pres/20021223a.html.

43. http://archiver.rootsweb.ancestry.com/th/read/GENEALOGY-DNA/2008 -02/1203878218.

44. http://www.scheidecker.net/personal-genome-explorer/.

45. K. Offit, "Genomic profiles for disease risk: Predictive or premature?" *Journal of the American Medical Association*, 2008, 299(11): 1353–55; "Direct-to-consumer genetic tests: Flawed and unethical," *Lancet Oncology*, 2008, 9(12): 1113; H. J. Bandelt, Y. G. Yao, M. B. Richards, and A. Salas, "The brave new era of human genetic testing," *Bioessays,* 2008, 30(11–12): 1246–51; W. G. Feero, A. E. Guttmacher, and F. S. Collins, "The genome gets personal—almost," *Journal of the American Medical Association*, 2008, 299(11): 1351–52.

46. G. B. van Ommen and M. C. Cornel, "Recreational genomics? Dreams and

fears on genetic susceptibility screening," *European Journal of Human Genetics*, 2008, 16(4): 403–4.

47. Email from Linda Avey, March 16, 2009.

48. D. J. Hunter, M. J. Khoury, and J. M. Drazen, "Letting the genome out of the bottle—will we get our wish?" *New England Journal of Medicine*, 2008, 358(2): 105–7.

49. http://oba.od.nih.gov/oba/SACGHS/meetings/july2008/transcripts/08JUL fullday.pdf.

50. http://meetings.cshl.edu/meetings/pastprograms/2008%20Past%20Pro grams/Genome%20Program.pdf.

51. http://oba.od.nih.gov/SACGHS/sacghs_past_meeting_2008_jul_08.html.

52. K. Seligman, "The social entrepreneur," *San Francisco Chronicle*, January 8, 2006.

53. Interview with Ryan Phelan, December 5, 2007.

54. Ibid.

55. http://talk.dnadirect.com/2008/06/14/dna-direct-in-full-compliance-with-state-and-national-regulations/.

56. http://oba.od.nih.gov/oba/SACGHS/meetings/july2008/transcripts/08JUL fullday.pdf.

57. Ibid.

58. Jim Evans at SACGHS meeting, July 8, 2008.

59. http://oba.od.nih.gov/oba/SACGHS/meetings/july2008/transcripts/08JUL fullday.pdf.

60. http://genomecommons.org/background/.

61. S. Levy, G. Sutton, P. C. Ng, et al., "The diploid genome sequence of an individual human," *PLoS Biology*, 2007, 5(10): e254.

62. Conversation with Steve Brenner, June 10, 2008.

63. Conversation with Hugh Rienhoff and Linda Avey, June 10, 2008.

64. G. Naik and A. Regalado, "A fitness mogul, stricken by illness, hunts for genes," *Wall Street Journal*, November 30, 2006.

65. http://www.alsforums.com/facts.php.

66. Naik and Regalado, "A fitness mogul."

67. Email from Dietrich Stephan, February 19, 2008.

68. Interview with Linda Avey, December 5, 2007.

69. Email from Dietrich Stephan, February 19, 2008.

70. Interview with George Church, December 6, 2007.

71. http://www.knome.com/about/news/20071127.html.

72. Interview with Stacey Gabriel, February 7, 2008.

73. Interview with Pardis Sabeti, April 15, 2008.

74. Interview with Tara Matise, April 15, 2008.

75. Email from Jason Bobe, March 1, 2008.

CHAPTER 5: BETTER LIVING THROUGH CHEMISTRY

1. C. A. Hutchison, "DNA sequencing: Bench to bedside and beyond," *Nucleic Acids Research*, 2007, 35(18): 6227–37.

2. Interview with Walter Gilbert, January 16, 2007.

3. G. L. Trainor, "DNA sequencing, automation, and the human genome," *Analytical Chemistry*, 1990, 62(5): 418–26.

4. B. L. Karger and A. Guttman, "DNA sequencing by CE," *Electrophoresis*, 2009, 30 Suppl 1: S196–202.

5. http://findarticles.com/p/articles/mi_hb6656/is_21_8/ai_n28763368/.

6. J. Shreeve, *The Genome War: How Craig Venter Tried to Capture the Code of Life and Save the World* (New York: Knopf, 2004).

7. Interview with Steve Lombardi, March 11, 2008.

8. Ibid.

9. Interview with Andy Watson, September 16, 2009.

10. http://hum-molgen.org/NewsGen/09-2003/000037.html.

11. http://www.genome.gov/12513210.

12. http://genomics.xprize.org/press-release/x-prize-foundation-announces-largest-medical-prize-in-history.

13. Interview with George Church, December 6, 2007.

14. Ibid.

15. Interview with Ian Goodhead, February 8, 2007.

16. K. Davies, "454's Rothberg speaks about 'sequencing by synthesis,'" *Bio-IT World*, August 18, 2005.

17. J. M. Rothberg and J. H. Leamon, "The development and impact of 454 sequencing," *Nature Biotechnology*, 2008, 26(10): 1117–24.

18. M. Margulies, M. Egholm, W. E. Altman, et al., "Genome sequencing in microfabricated high-density picolitre reactors," *Nature*, 2005, 437(7057): 376–80.

19. M. L. Metzker, "Sequencing technologies—the next generation," *Nature Reviews Genetics*, 2010, 11(1): 31–46.

20. Margulies, Egholm, Altman, et al., "Genome sequencing in microfabricated high-density picolitre reactors."

21. http://www.genengnews.com/articles/chitem.aspx?aid=658&chid=2.

22. K. Davies, "454's Rothberg speaks about 'sequencing by synthesis.'"

23. http://www.454.com/about-454/news/index.asp?display=detail&id=66.

24. Interview with Chad Nusbaum, October 21, 2008.

25. http://www.pbs.org/wgbh/nova/genome/deco_lander.html.

26. E. Pennisi, "Genomics: Semiconductors inspire new sequencing technologies." *Science*, 2010, 327(5970): 1190.

27. Interview with Stan Lapidus, March 11, 2008.

28. L. Braslavsky, B. Hebert, E. Kartalov, and S. R. Quake, "Sequence information can be obtained from single DNA molecules," *Proceedings of the National Academy of Sciences*, 2003, 100(7): 3960–64.

29. Interview with Stan Lapidus, March 11, 2008.

30. K. B. Mullis, "The unusual origin of the polymerase chain reaction," *Scientific American*, 1990, 262(4): 56–61, 64–65.

31. Y. M. Lo and K. C. Chan, "Introduction to the polymerase chain reaction," *Methods in Molecular Biology*, 2006, 336: 1–10.

32. K. B. Mullis, *Dancing Naked in the Mind Field*, 1st ed. (New York: Pantheon, 1998).

33. Interview with Jay Shendure, October 28, 2006.

34. Interview with Stan Lapidus, March 11, 2008.

35. Ibid.

36. Ibid.

37. http://ir.helicosbio.com/releasedetail.cfm?ReleaseID=236436.

38. Interview with Tim Harris, October 22, 2009.

39. http://ir.helicosbio.com/scientific_advisory.cfm.

40. http://ir.helicosbio.com/committees.cfm?bio=15416.

41. Interview with Stan Lapidus, March 11, 2008.

42. Interview with Steve Lombardi, October 26, 2006.

43. H. Auer, D. L. Newsom, and K. Kornacker, "Expression profiling using Affymetrix genechip microarrays," *Methods in Molecular Biology*, 2009, 509: 35–46.

44. G. G. Marcial, "Fast growth at Illumina is in the genes," *BusinessWeek*, February 20, 2006.

45. "New chips on the block; case history," *Economist*, December 2, 2006.

46. Interview with Steve Lombardi, October 26, 2006.

47. Ibid.

48. Ibid.

49. http://www.genengnews.com/news/bnitem.aspx?name=14953823.

50. C. Shaffer, "Next-generation sequencing outpaces expectations," *Nature Biotechnology*, 2007, 25(2): 149.

51. http://google.brand.edgar-online.com/EFX_dll/EDGARpro.dll?FetchFiling HTML1?SessionID=QthBWO6BpJMDMKO&ID=5353168-44499-73147.

52. Interview with Stan Lapidus, March 11, 2008.

53. Ibid.

54. Interview with John Sullivan, March 10, 2008.

55. Interview with Un Kwon-Casado, December 6, 2007.

56. Interview with Steve Lombardi, February 8, 2007.

57. Interview with Tim Harris, October 22, 2009.

58. Interview with Kevin Shianna, February 9, 2007.

59. Interview with Un Kwon-Casado, December 6, 2007.

60. Ibid.

61. Interview with Chad Nusbaum, February 8, 2008.

62. Interview with Un Kwon-Casado, December 6, 2007.

63. Interview with Aravinda Chakravarti, February 7, 2007.

CHAPTER 6: AND THEN THERE WERE TEN

1. http://www.edventure.com.

2. http://www.ned.org/about/about.html.

3. Interview with Esther Dyson, September 7, 2007.

4. http://www.edge.org/documents/digerati/Dyson.html.

5. http://www.charlierose.com/view/interview/8640.

6. http://www.huffingtonpost.com/esther-dyson/release-09-knowledge-forc _b_61594.html.

7. http://online.wsj.com/article/SB118532736853177075.html.

8. Interview with Esther Dyson, September 7, 2007.

9. Ibid.

10. Interview with Esther Dyson, October 20, 2008.

11. Interview with Ting Wu, July 17, 2007.

12. Interview with Marie Wu, July 17, 2007.

13. http://www.personalgenomes.org/public/5.html.

14. http://www.cabrimed.org/book.jsp.

15. Interview with Kirk Maxey, July 18, 2007.

16. http://www.pbs.org/wnet/religionandethics/week952/feature.html.

17. Interview with Kirk Maxey, July 18, 2007.

18. Ibid.

19. Ibid.

20. http://news.bbc.co.uk/onthisday/hi/dates/stories/july/25/newsid_249000/2499411.stm.

21. J. Cohen, A. Trounson, K. Dawson, et al., "The early days of IVF outside the UK," *Human Reproduction Update*, 2005, 11(5): 439–59.

22. Interview with Kirk Maxey, July 18, 2007.

23. J. V. Linden and G. Centola, "New American Association of Tissue Banks standards for semen banking," *Fertility & Sterility*, 1997, 68(4): 597–600.

24. D. E. Pegg, "The history and principles of cryopreservation," *Seminars in Reproductive Medicine*, 2002, 20(1): 5–13.

25. Interview with Kirk Maxey, July 18, 2007.

26. Ibid.

27. http://transcripts.cnn.com/TRANSCRIPTS/0511/27/sm.02.html.

28. M. Booth, "Boy wonder," *Denver Post*, November 28, 2004.

29. http://www.donorsiblingregistry.com/about.php.

30. A. Motluk, "Anonymous sperm donor traced on Internet," *New Scientist*, November 3, 2005; S. Kotler, "The god of sperm," *LA Weekly*, September 27, 2007.

31. Interview with Kirk Maxey, July 18, 2007.

32. Email from Kirk Maxey, December 2, 2008.

33. Interview with Kirk Maxey, July 17, 2007.

34. Ibid.

35. http://www.cabrimed.org/donorconceivedservices.jsp.

36. Interview with Kirk Maxey, July 17, 2007.

37. http://www.fda.gov/BiologicsBloodVaccines/GuidanceComplianceRegulatoryInformation/Guidances/Tissue/ucm073964.htm.

38. http://www.aatb.org/AF_accreditedBank.asp.

39. S. Kotler, "The god of sperm."

40. A. Motluk, "Who's the daddy? US sperm banks must be better regulated," *New Scientist*, August 9, 2007.

41. C. Dyer, "Experts suggest ways to tackle shortage of sperm donors," *BMJ*, 2009, 338: b2620.

42. E. Ekerhovd, A. Faurskov, and C. Werner, "Swedish sperm donors are driven by altruism, but shortage of sperm donors leads to reproductive travelling," *Upsala Journal of Medical Sciences*, 2008, 113(3): 305–13.

43. Email from Kirk Maxey, December 2, 2008.

44. S. Kotler, "The god of sperm"; J. Wolff, "The truth about donor 1084," *SELF*, October 2006.

45. http://www.spermbankinformation.com/2008/07/27/national-sperm-donor-registry/.

46. http://www.pbs.org/wnet/religionandethics/week952/feature.html.

47. K. Maxey, *Pig Blood* (Ann Arbor, Mich.: Cayman Biomedical Research Institute, 2006).

48. R. Lehmann-Haupt, "Mapping the god of sperm," *Newsweek*, December 16, 2009.

49. Email from Kirk Maxey, December 1, 2008.

50. Email from Kirk Maxey, December 2, 2008.

51. Ibid.

52. Ibid.

53. Interview with George Church, June 6, 2007.

54. Email from George Church to the PGP-10, May 14, 2007.

55. Interview with George Church, June 6, 2007.

56. Ibid.

57. Ibid.

58. Ibid.

59. Email from George Church to the PGP-10, May 14, 2007.

60. Interview with George Church, March 10, 2008.

61. http://www.improb.com/projects/hair/hair-club-top.html.

62. pinker.wjh.harvard.edu/articles/media/2004_08_salon.pdf.

63. J. Crace, "Steven Pinker: The evolutionary man," guardian.co.uk, June 17, 2008.

64. pinker.wjh.harvard.edu/ . . . /dialogue%20Pinker%20and%20McEwan.pdf.

65. S. Paulson, "Proud atheists," Salon.com, October 15, 2007.

66. Interview with Steve Pinker, October 20, 2008.

67. S. Pinker, "My genome, my self," *New York Times Magazine*, January 7, 2009.

68. Interview with Steve Pinker, October 20, 2008.

69. Interview with Keith Batchelder, July 17, 2007.

70. http://www.personalgenomes.org/public/7.html.

71. Interview with Keith Batchelder, July 17, 2007.

72. Ibid.

73. Ibid.

74. First annual meeting of Personal Genome Project participants, July 17, 2007.

75. Interview with Keith Batchelder, July 17, 2007.

76. L. M. Hynicka, W. D. Cahoon, Jr., and B. L. Bukaveckas, "Genetic testing for warfarin therapy initiation," *Annals of Pharmacotherapy*, 2008, 42(9): 1298–303.

77. Interview with Keith Batchelder, July 17, 2007.

78. L. J. Lesko, "The critical path of warfarin dosing: Finding an optimal dosing strategy using pharmacogenetics," *Clinical Pharmacology & Therapeutics*, 2008, 84(3): 301–3.

79. Ibid.

80. Interview with Keith Batchelder, July 17, 2007.

81. Interview with Stan Lapidus, March 11, 2008.

82. Ibid.

83. Ibid.

84. Interview with Jorge Conde, October 20, 2008.

85. Interview with Stan Lapidus, March 11, 2008.

86. Ibid.

87. counsyl.com.

88. Interview with Stan Lapidus, March 11, 2008.

89. V. Nossov, M. Amneus, F. Su, et al., "The early detection of ovarian cancer: From traditional methods to proteomics. Can we really do better than serum ca-125?" *American Journal of Obstetrics & Gynecology*, 2008, 199(3): 215–23; M. Argento, P. Hoffman, and A. S. Gauchez, "Ovarian cancer detection and treatment: Current situation and future prospects," *Anticancer Research*, 2008, 28(5B): 3135–38.

90. A. Pollack, "Cancer test for women raises hope, and concern," *New York Times*, August 26, 2008; I. Visintin, Z. Feng, G. Longton, et al., "Diagnostic markers for early detection of ovarian cancer," *Clinical Cancer Research*, 2008, 14(4): 1065–72.

91. D. Levenson, "The trouble with OvaSure," *American Journal of Medical Genetics: Part A,* 2009, 149A(3): viii–ix.

92. http://therumpus.net/2009/01/what-happened-to-sheila/.

93. Interview with Rosalynn Gill-Garrison, 20 October 2008.

94. www.casewatch.org/hearings/dna.pdf.

95. Interview with Rosalynn Gill-Garrison, October 20, 2008.

96. http://gao.gov/products/GAO-06-977T.

97. Interview with Rosalynn Gill-Garrison, October 20, 2008.

98. M. Ritter, "Looking for a healthier diet? Testing your DNA may help," Associated Press, October 14, 2005.

99. www.casewatch.org/hearings/dna.pdf.

100. Interview with Rosalynn Gill-Garrison, October 20, 2008.

101. Ibid.

102. Ibid.

103. http://gao.gov/products/GAO-06-977T.

104. http://www.hgc.gov.uk/Client/document.asp?DocId=94&CAtegoryId=3.

105. Comment by Rosalynn Gill-Garrison at the first Personal Genome Project gathering, July 17, 2007.

106. http://www.aboutus.org/Mycellf.com.

107. Email from Rosalynn Gill-Garrison, September 28, 2009.

108. http://www.ipsogen.com/corporate/science-technologies/.

109. P. A. Lombardo and G. M. Dorr, "Eugenics, medical education, and the public health service: Another perspective on the Tuskegee Syphilis Experiment," *Bulletin of the History of Medicine*, 2006, 80(2): 291–316.

110. C. Holden, "Air Force challenged on sickle trait policy," *Science*, 1981, 211(4479): 257.

111. J. Bussey-Jones, G. Henderson, J. Garrett, M. Moloney, C. Blumenthal, and G. Corbie-Smith, "Asking the right questions: Views on genetic variation research among black and white research participants," *Journal of General Internal Medicine*, 2009, 24(3): 299–304; P. Achter, R. Parrott, and K. Silk, "African Americans' opinions about human-genetics research," *Politics and the Life Sciences*, 2004, 23(1): 60–66.

112. James Sherley comment at the second annual gathering of the Personal Genome Project, October 20, 2008.

113. R. Skloot, *The Immortal Life of Henrietta Lacks* (New York: Crown Publishers, 2010).

114. Interview with James Sherley, July 17, 2007.

115. Ibid.

116. Interview with James Sherley, March 9, 2008.

117. web.mit.edu/fnl/volume/sherley/chomskyetal_sherley.pdf.

118. J. Wolfson, "When race enters the equation," *Boston Magazine*, July 2007.

119. http://stemcells.nih.gov/policy/2001policy.htm.

120. C. Kalb and D. Rosenberg, "Stem cell division," *Newsweek*, October 25, 2004.

121. http://www.npr.org/templates/story/story.php?storyId=5252449.

122. clmagazine.org/backissues/2007janfeb_20-23jamessherley.pdf.

123. Interview with James Sherley, March 9, 2008.

124. http://nihroadmap.nih.gov/pioneer/Recipients06.aspx.

125. J. Kwan, "BE professor threatens hunger strike to protest tenure denial," *The Tech*, January 10, 2007.

126. http://web.mit.edu/provost/letters/letter05142007.html.

127. Interview with James Sherley, March 9, 2008.

128. http://tech.mit.edu/V127/N1/1facultyopn.html.

129. http://web.mit.edu/newsoffice/2007/statements-sherley.html.

130. http://web.mit.edu/fnl/volume/sherley/BE_sherley(3.30).pdf.

131. http://pgen.us/Sherley/16._May_14.facultye-mail.Reif.pdf.

132. http://pgen.us/April_17.ManningLetter.pdf.

133. pgen.us/May15Tech-Reif.pdf.

134. http://chronicle.com/article/Scientist-in-Tenure-Fight-With/39151/.

135. J. Kwan, "Prof. Sherley locked out of BE laboratory after June 30 deadline passes," *The Tech*, July 6, 2007.

136. Interview with James Sherley, July 18, 2007.

137. http://www.thecrimson.com/article/2007/2/21/on-strike-in-1932-mahatma-gandhi/.

138. Interview with James Sherley, July 18, 2007.

139. http://m.insidehighered.com/news/2008/03/21/mit.

140. Interview with James Sherley, March 9, 2008.

141. Interview with James Sherley, October 20, 2008.

142. http://web.mit.edu/provost/raceinitiative/toc.html.

143. Interview with James Sherley, July 17, 2007.

144. Ibid.

145. Interview with James Sherley, March 9, 2008.

146. T. Graf and T. Enver, "Forcing cells to change lineages," *Nature*, 2009, 462(7273): 587–94.

147. H. S. Lee, G. G. Crane, J. R. Merok, et al., "Clonal expansion of adult rat hepatic stem cell lines by suppression of asymmetric cell kinetics (sack)," *Biotechnology & Bioengineering*, 2003, 83(7): 760–71; L. Rambhatla, S. Ram-Mohan, J. J. Cheng, and J. L. Sherley, "Immortal DNA strand cosegregation requires p53/IMPDH-dependent asymmetric self-renewal associated with adult stem cells," *Cancer Research*, 2005, 65(8): 3155–61; J. L. Sherley, P. B. Stadler, and D. R. Johnson, "Expression of the wild-type p53 antioncogene induces guanine nucleotide-dependent stem cell division kinetics," *Proceedings of the National Academy of Sciences*, 1995, 92(1): 136–40.

148. Interview with James Sherley, March 9, 2008.

149. Ibid.

150. Interview with George Church, October 26, 2006.

151. Interview with James Sherley, March 9, 2008.

152. Ibid.

CHAPTER 7: "IT'S TOUGH TO GUARD AGAINST THE FUTURE"

1. E. R. Mardis, "Next-generation DNA sequencing methods," *Annual Review of Genomics & Human Genetics*, 2008, 9: 387–402.

2. A. Coombs, "The sequencing shakeup," *Nature Biotechnology*, 2008, 26(10): 1109–12.

3. http://www.polonator.org/.

4. Interview with George Church, October 10, 2007.

5. J. Shreeve, *The Genome War: How Craig Venter Tried to Capture the Code of Life and Save the World* (New York: Knopf, 2004).

6. Interview with Chad Nusbaum, February 8, 2008.

7. Interview with Kevin McCarthy, February 6, 2008.

8. Ibid.

9. Amy McGuire presentation at AGBT, Marco Island, Florida, February 8, 2008.

10. H. Bickeboller, D. Campion, A. Brice et al., "Apolipoprotein E and Alzheimer disease: Genotype-specific risks by age and sex," *American Journal of Human Genetics*, 1997, 60(2): 439–46.

11. Interview with James Watson, May 11, 2007.

12. Previously at http://ep2008.europython.eu/Talks%20and%20Themes/Speakers.

13. Mike Cariaso, AGBT meeting, Marco Island, Florida, February 8, 2008.

14. D. R. Nyholt, C. E. Yu, and P. M. Visscher, "On Jim Watson's APOE status: Genetic information is hard to hide," *European Journal of Human Genetics*, 2009, 17(2): 147–49.

15. Email from Mike Cariaso, February 17, 2008.

16. http://www.snpedia.com/index.php/Rs6457617.

17. A. Barton and J. Worthington, "Genetic susceptibility to rheumatoid arthritis: An emerging picture," *Arthritis & Rheumatism*, 2009, 61(10): 1441–46.

18. Interview with Greg Lennon, December 17, 2008.

19. Ibid.

20. G. Hardiman, "Microarray platforms—comparisons and contrasts," *Pharmacogenomics*, 2004, 5(5): 487–502.

21. Interview with Greg Lennon, December 17, 2008.

22. Email from Mike Cariaso, June 1, 2009.

23. Interview with Greg Lennon, December 17, 2008.

24. Ibid.

25. www.ncbi.nlm.nih.gov/omim/.

26. V. A. McKusick, *Mendelian Inheritance in Man: Catalogs of Autosomal Dominant,*

Autosomal Recessive, and X-Linked Phenotypes (Baltimore: Johns Hopkins University Press, 1966).

27. V. A. McKusick, "*Mendelian Inheritance in Man* and its online version, OMIM," *American Journal of Human Genetics*, 2007, 80(4): 588–604.

28. http://www.guardian.co.uk/uk/2004/feb/10/booksnews.ireland.

29. http://www.ncbi.nlm.nih.gov/projects/SNP/snp_summary.cgi.

30. Interview with Greg Lennon, December 17, 2008.

31. Ibid.

32. http://www.lyricsmania.com/therapy_lyrics_loudon_wainwright.html.

33. Email exchange with Mike Cariaso, March 2, 2008.

34. Conversation with Dietrich Stephan, March 25, 2008.

35. A. L. McGuire and W. Burke, "An unwelcome side effect of direct-to-consumer personal genome testing: Raiding the medical commons," *JAMA*, 2008, 300(22): 2669–71.

36. T. Ray, "Navigenics lowers Health Compass price to $1K," *Pharmacogenomics*, July 22, 2009.

CHAPTER 8: GETTYSBURG TO GUTENBERG

1. Interview with Robert Green, March 11, 2008.

2. D. F. Ransohoff and M. J. Khoury, "Personal genomics: Information can be harmful," *European Journal of Clinical Investigation*, 2010, 40(1): 64–68.

3. R. C. Green, J. S. Roberts, L. A. Cupples, et al., "Disclosure of APOE genotype for risk of Alzheimer's disease," *New England Journal of Medicine*, 2009, 361(3): 245–54.

4. Interview with Robert Green, March 11, 2008.

5. Ibid.

6. Interview with George Church, October 3, 2008.

7. Interview with Robert Green, March 11, 2008.

8. Interview with George Church, February 21, 2008.

9. http://arep.med.harvard.edu/P50_03/.

10. http://arep.med.harvard.edu/PGP/Anon.htm.

11. Interview with Jeff Schloss, January 5, 2007.

12. Interview with George Church, March 13, 2007.

13. Interview with Terry Bard, October 26, 2006.

14. http://www.genome.gov/20519355.

15. Interview with Les Biesecker, September 21, 2007.

16. Ibid.

17. L.G. Biesecker, J.C. Mullikin, F.M. Facio, et al., "The ClinSeq project: Piloting large-scale genome sequencing for research in genomic medicine." *Genome Research*, 2009, 19(9): 1665–74.

18. Interview with Les Biesecker, September 21, 2007.

19. Amy McGuire presentation at the National Institute of Environmental Health Sciences, June 15, 2007.

20. Interview with Ting Wu, March 10, 2008.

21. Interview with Robert Green, March 11, 2008.

22. Interview with Jason Bobe, February 27, 2008.

23. Interview with Gail Henderson, March 27, 2008.

24. Ibid.

25. Ibid.

26. Ibid.

27. Interview with Chad Nusbaum, March 2008.

28. Interview with Jeantine Lunshof, March 4, 2008.

29. Interview with Jason Bobe, March 1, 2008.

30. Interview with Robert Green, March 11, 2008.

31. Interview with George Church, March 10, 2008.

32. Interview with James Sherley, March 9, 2008.

33. Interview with Chad Nusbaum, October 21, 2008.

34. Email from George Church, December 30, 2008.

35. grants.nih.gov/grants/peer/ . . . /scoring_system_and_procedure.pdf.

36. Email from George Church, December 18, 2008.

37. Summary Statement, NHGRI Special Emphasis Panel, Human Gene/Environment Trait Technology Center, December 1, 2008, e-mailed to the author by George Church.

38. http://en.wikipedia.org/wiki/Synchronicity_II.

39. Email from George Church, December 30, 2008.

40. Ibid.

41. Email to NHGRI administrators, February 2, 2009.

42. Interview with George Church, July 18, 2009.

43. Sixteenth Meeting of the Secretary's Advisory Committee on Genetics, Health and Society, July 7–8, 2008.

44. Interview with George Church, August 11, 2008.

45. T. Ray, "Navigenics agrees not to market genetic testing services directly to NY residents," *Pharmacogenomics Reporter*, January 14, 2010.

CHAPTER 9: "YOU CAN DO THIS IN YOUR KITCHEN"

1. http://io9.com/338347/addicted-to-the-future.
2. http://partsregistry.org/Main_Page.
3. http://bbf.openwetware.org/FAQ.html.
4. http://io9.com/5022316/mad-science-contest-build-a-lifeform-and-well-send-you-to-hong-kong-or-give-you-1000.
5. Interview with Kay Aull, June 8, 2009.
6. http://io9.com/5049788/making-a-biological-counter.
7. Ibid.
8. M. A. Shampo and R. A. Kyle, "Kary B. Mullis—Nobel laureate for procedure to replicate DNA," *Mayo Clinic Proceedings*, 2002, 77(7): 606.
9. M. F. Kramer and D. M. Coen, "Enzymatic amplification of DNA by PCR: Standard procedures and optimization," *Current Protocols in Cytometry*, August 2006, Appendix 3K.
10. Interview with Kay Aull, June 8, 2009.
11. http://diybio.org/about/.
12. Interview with Kay Aull, June 8, 2009.
13. T. Wallack, "Codon Devices closing as financing dwindles," *Boston Globe*, April 3, 2009.
14. Interview with Kay Aull, June 8, 2009.
15. Ibid.
16. M. C. Janssen and D. W. Swinkels, "Hereditary haemochromatosis," *Best Practice & Research Clinical Gastroenterology*, 2009, 23(2): 171–83.
17. Interview with Kay Aull, June 8, 2009.
18. Ibid.
19. http://www.360.monitor.com/.
20. J. Whalen, "In attics and closets, 'biohackers' discover their inner Frankenstein," *Wall Street Journal*, May 12, 2009.
21. http://groups.google.com/group/diybio/msg/060712ed9c7ffb5c.
22. Interview with Kay Aull, June 8, 2009.
23. R. Hirsch, "The strange case of Steve Kurtz: Critical Art Ensemble and the price of freedom," *Afterimage*, May–June 2005: 22–32.
24. S. Kurtz, "The four-year fight for biological art. Steven Kurtz interviewed by Rachel Courtland," *Nature*, 2008, 453(7196): 707.
25. Interview with Kay Aull, June 8, 2009.
26. http://bioweathermap.org/about.html.
27. http://bioweathermap.org/.

28. Interview with Kay Aull, June 8, 2009.

29. Email from Kay Aull, October 14, 2009.

30. Email from Kay Aull, January 18, 2010.

31. Email from Kay Aull, January 19, 2010.

32. Interview with Hugh Rienhoff, December 9, 2008.

33. Interview with Hugh Rienhoff, May 20, 2008.

34. Ibid.

35. R. Maymon and A. Herman, "The clinical evaluation and pregnancy outcome of euploid fetuses with increased nuchal translucency," *Clinical Genetics*, 2004, 66(5): 426–36.

36. Interview with Hugh Rienhoff, May 20, 2008.

37. http://www.marfan.org/marfan/2319/Diagnosis.

38. C. Kroen, "Abraham Lincoln and the 'Lincoln sign,'" *Cleveland Clinic Journal of Medicine*, 2007, 74(2): 108–10.

39. N. M. Ammash, T. M. Sundt, and H. M. Connolly, "Marfan syndrome—diagnosis and management," *Current Problems in Cardiology*, 2008, 33(1): 7–39.

40. Interview with Hugh Rienhoff, May 20, 2008.

41. Interview with Lisa Hane, December 9, 2008.

42. B. L. Callewaert, B. L. Loeys, A. Ficcadenti, et al., "Comprehensive clinical and molecular assessment of 32 probands with congenital contractural arachnodactyly: Report of 14 novel mutations and review of the literature," *Human Mutation*, 2009, 30(3): 334–41.

43. Email from Hugh Rienhoff, August 7, 2009.

44. Interview with Hugh Rienhoff, May 20, 2008.

45. Ibid.

46. http://www.ncbi.nlm.nih.gov/pmc/articles/PMC2227915/.

47. P. A. Reichart and H. P. Philipsen, *Oral Pathology: Color Atlas of Dental Medicine*, ed. K. H. Rateitschak and H. F. Wolf (New York: Thieme, 2000).

48. Interview with Hugh Rienhoff, May 20, 2008.

49. B. L. Loeys, U. Schwarze, T. Holm, et al., "Aneurysm syndromes caused by mutations in the TGF-beta receptor," *New England Journal of Medicine*, 2006, 355(8): 788–98.

50. Interview with Hugh Rienhoff, May 20, 2008.

51. H. D. Kollias and J. C. McDermott, "Transforming growth factor–beta and myostatin signaling in skeletal muscle," *Journal of Applied Physiology*, 2008, 104(3): 579–87.

52. D. Martindale, "Muscle twitch switch," *Scientific American*, 2004, 291(6): 22, 24.

53. E. M. McNally, "Powerful genes—myostatin regulation of human muscle mass," *New England Journal of Medicine*, 2004, 350(26): 2642–44.

54. Interview with Hugh Rienhoff, May 20, 2008.

55. A. C. McPherron, A. M. Lawler, and S. J. Lee, "Regulation of skeletal muscle mass in mice by a new TGF-beta superfamily member," *Nature*, 1997, 387(6628): 83–90.

56. Interview with Hugh Rienhoff, May 20, 2008.

57. Ibid.

58. Ibid.

59. P. Matt, J. Habashi, T. Carrel, D. E. Cameron, J. E. Van Eyk, and H. C. Dietz, "Recent advances in understanding Marfan syndrome: Should we now treat surgical patients with losartan?" *Journal of Thoracic & Cardiovascular Surgery*, 2008, 135(2): 389–94.

60. Interview with Hal Dietz, December 18, 2008.

61. E. R. Neptune, P. A. Frischmeyer, D. E. Arking, et al., "Dysregulation of TGF-beta activation contributes to pathogenesis in Marfan syndrome," *Nature Genetics*, 2003, 33(3): 407–11.

62. Interview with Hal Dietz, July 7, 2008.

63. J. Travis, "Medicine. Old drug, new hope for Marfan syndrome," *Science*, 2006, 312(5770): 36–37.

64. Interview with Hal Dietz, December 18, 2008.

65. Interview with Hugh Rienhoff, May 20, 2008.

66. Ibid.

67. Interview with Hugh Rienhoff, December 7, 2008.

68. Ibid.

69. Ibid.

70. http://www.marfan.org/marfan/2626/APR—Stanford-University.

71. B. Koerner, "Decoding Beatrice: A father's quest," *Wired*, January 19, 2009.

72. Interview with Hugh Rienhoff, December 7, 2008.

73. Ibid.

74. B. Koerner, "Decoding Beatrice: A father's quest."

75. Interview with Hugh Rienhoff, December 7, 2008.

76. "Within spitting distance," Economist.com, November 20, 2007.

77. B. Maher, "Personal genomics: His daughter's DNA," *Nature*, 2007, 449(7164): 773–76.

78. H. Y. Rienhoff, "My daughter's DNA," *Make*, December 17, 2008.

79. B. Koerner, "Decoding Beatrice: A father's quest."

80. http://www.youtube.com/watch?v=4WOaQhjWmRU.

81. http://mydaughtersdna.org.

82. Interview with Hugh Rienhoff, December 7, 2008.

83. Interview with Hugh Rienhoff, May 20, 2008.

84. Interview with Hal Dietz, July 7, 2008.

85. Interview with Hugh Rienhoff, December 7, 2008.

86. Interview with Hugh Rienhoff, August 6, 2009.

87. Ibid.

88. Ibid.

CHAPTER 10: "TAKE A CHANCE, WIN A BUNNY"

1. Interview with Sarah Angrist, June 15, 2007.

2. J. De Greve, E. Sermijn, S. De Brakeleer, Z. Ren, and E. Teugels, "Hereditary breast cancer: From bench to bedside," *Current Opinion in Oncology*, 2008, 20(6): 605–13.

3. W. Krause, "Male breast cancer—an andrological disease: Risk factors and diagnosis," *Andrologia*, 2004, 36(6): 346–54.

4. C. Schubert, "Mary-Claire King," *Nature Medicine*, 2003, 9(6): 633.

5. K. Davies and M. White, *Breakthrough: The Race to Find the Breast Cancer Gene* (New York: Wiley, 1996).

6. T. Walsh, S. Casadei, K. H. Coats, et al., "Spectrum of mutations in BRCA1, BRCA2, CHEK2, and TP53 in families at high risk of breast cancer," *JAMA*, 2006, 295(12): 1379–88.

7. Conversation with Mary-Claire King, October 12, 2008.

8. Interview with Rich Terry, March 10, 2008.

9. Interviews with Rich Terry and Greg Porreca, March 10, 2008.

10. Interview with George Church, March 10, 2008.

11. http://www.google.com/patents/about?id=QeejAAAAEBAJ&dq=polony&num=100.

12. J. Karow, "For ABI, developing Agencourt's sequencing technology is a high priority," *GenomeWeb Daily News*, June 6, 2006.

13. K. Davies, "The drive for the $1000 genome," *Bio-IT World*, May 15, 2007.

14. http://www.equitygroups.com/nyse/abi/messages/144027.html.

15. Interview with Kevin McKernan, February 8, 2007.

16. Interview with Chad Nusbaum, February 2, 2009.

17. Interview with George Church, July 18, 2007.

18. Interview with George Church, December 5, 2008.

19. Interview with George Church, March 16, 2009.

20. Interview with Chad Nusbaum, February 2, 2009

21. Interview with Kevin McCarthy, January 2010.

22. Interview with Greg Porreca, April 16, 2009.

23. Interview with Kevin McCarthy, January 20, 2010.

24. Email from George Church, December 14, 2009.

25. http://ir.helicosbio.com/releasedetail.cfm?ReleaseID=438340.

26. Interview with George Church, January 18, 2010.

27. E. Winnick and J. Karow, "Helicos sees five to 10 orders by end '08, shows resequencing, gene-expression results," *In Sequence*, September 30, 2008.

28. "Helicos to lay off 30 percent of workforce to cut costs; retracts order guidance," *GenomeWeb Daily News*, December 4, 2008.

29. L. Timmerman, "Helicos shuffles CEOs, names Lowy to top job," *Xconomy*, December 2, 2008.

30. Interview with Steve Lombardi, December 14, 2009.

31. "Expression Analysis returns Helicos sequencing system," *GenomeWeb Daily News*, January 30, 2009.

32. Interview with Steve McPhail, January 23, 2009.

33. Anonymous, "Expression analysis certified on Illumina platforms," *GenomeWeb Daily News*, February 15, 2010.

34. J. Karow, "Helicos installs two instruments in Q2, seeks new funding sources as $5M of cash remain," *In Sequence*, August 18, 2009.

35. http://ir.helicosbio.com/releasedetail.cfm?ReleaseID=406880.

36. Interview with Chad Nusbaum, October 21, 2008.

37. T. D. Harris, P. R. Buzby, H. Babcock, et al., "Single-molecule DNA sequencing of a viral genome," *Science*, 2008, 320(5872): 106–9.

38. http://www.genomeweb.com/sequencing/tim-harris-francis-collins.

39. Interview with Tim Harris, October 22, 2009.

40. "Helicos to raise $9.4M in private placement," *GenomeWeb Daily News*, September 16, 2009.

41. http://www.genomeweb.com/eight-firms-gwdn-index-return-triple-digit-stock-price-increases-09.

42. D. Pushkarev, N. F. Neff, and S. R. Quake, "Single-molecule sequencing of an individual human genome," *Nature Biotechnology*, 2009, 27(9): 847–52.

43. F. Ozsolak, A. R. Platt, D. R. Jones, et al., "Direct RNA sequencing," *Nature*, 2009, 461(7265): 814–18.

44. J. Karow, "High-throughput sequencing gains traction in disease RX in 09; new platforms to enter this year," *In Sequence*, January 5, 2010.

45. "Helicos sells four instruments to Riken," *GenomeWeb Daily News*, September 8, 2009.

46. K. Lakhman, "Correlagen debuts Helicos-based cardiac panel; is there room for Illumina's GA2?" *Sample*, October 19, 2009.

47. J. Karow, "High-throughput sequencing gains traction in disease RX in 09; new platforms to enter this year."

48. http://ir.helicosbio.com/secfiling.cfm?filingid=1104659-10-6957.

49. Interview with Steve Lombardi, February 16, 2010.

50. Anonymous, "Investment bank drops coverage of helicos biosciences," *GenomeWeb Daily News*, March 8, 2010.

51. B. Toner and J. Karow, "Helicos 'repositioning' as cash dwindles, institute returns HeliScope, Nasdaq threatens delisting," *In Sequence*, April 20, 2010.

52. P. Feugier and J. M. Chevalier, "The Paget-Schroetter syndrome," *Acta Chirurgica Belgica*, 2005, 105(3): 256–64.

53. D. Wardrop and D. Keeling, "The story of the discovery of heparin and warfarin," *British Journal of Haematology*, 2008, 141(6): 757–63.

54. D. Kurnik, R. Loebstein, H. Halkin, E. Gak, and S. Almog, "10 years of oral anticoagulant pharmacogenomics: What difference will it make? A critical appraisal," *Pharmacogenomics*, 2009, 10(12): 1955–65.

55. R. Weinshilboum and L. Wang, "Pharmacogenomics: Bench to bedside," *Nature Reviews: Drug Discovery*, 2004, 3(9): 739–48; R. Sikka, B. Magauran, A. Ulrich, and M. Shannon, "Bench to bedside: Pharmacogenomics, adverse drug interactions, and the cytochrome p450 system," *Academic Emergency Medicine*, 2005, 12(12): 1227–35; S. Sattiraju, S. Reyes, G. C. Kane, and A. Terzic, "K(ATP) channel pharmacogenomics: From bench to bedside," *Clinical Pharmacology & Therapeutics*, 2008, 83(2): 354–57.

56. A. Gawande, *The Checklist Manifesto: How to Get Things Right* (New York: Metropolitan, 2010); A. Gawande, "The checklist: If something so simple can transform intensive care, what else can it do?" *New Yorker*, 2007: 86–101.

57. J. Laurance, "Peter Pronovost: Champion of checklists in critical care," *Lancet*, 2009, 374(9688): 443.

58. E. Munoz, W. Munoz 3rd, and L. Wise, "National and surgical health care expenditures, 2005–2025," *Annals of Surgery*, 2010.

59. Interview with Marilyn Ness, October 19, 2008.

60. P. L. Hedley, P. Jorgensen, S. Schlamowitz, et al., "The genetic basis of Brugada syndrome: A mutation update," *Human Mutation*, 2009, 30(9): 1256–66.

61. Email from Joe Thakuria, February 4, 2010.

62. Interview with George Church, October 19, 2008.

63. Interview with George Church, Ting Wu, Joe Thankuria, Church's office, October 19, 2008.

64. Ibid.

65. Ibid.

66. Ibid.

67. Interview with George Church, October 19, 2008.

68. J. Zhou and E. A. Shephard, "Mutation, polymorphism and perspectives for the future of human flavin-containing monooxygenase 3," *Mutation Research*, 2006, 612(3): 165–71.

69. Email from Joe Thakuria, February 4, 2010.

70. Comments from participants in the second annual gathering of the Personal Genome Project, October 20, 2008.

71. M. W. Foster, J. J. Mulvihill, and R. R. Sharp, "Evaluating the utility of personal genomic information," *Genetics in Medicine*, 2009, 11(8): 570–74.

72. Comments from participants in the second annual gathering of the Personal Genome Project, October 20, 2008.

73. Email from Jason Bobe, January 7, 2009.

74. Interview with George Church, February 2, 2009.

75. Interview with Sasha Wait Zaranek, February 2, 2009.

76. Ibid.

77. Ibid.

78. Interview with George Church, October 20, 2008.

79. Ibid.

80. Conversation with Lisa Kessler, February 5, 2009.

81. Conversation with Lisa Kessler, February 26, 2009.

82. C. Tilstone, "Gene patents: A myriad of issues," *Lancet Oncology*, 2004, 5(7): 392.

83. E. Verkey, "Patenting of medical methods—need of the hour," *Journal of Intellectual Property Law & Practice*, 2006, 2(2): 104–13.

84. G. Poste, "The case for genomic patenting," *Nature*, 1995, 378(6557): 534–36.

85. R. Cook-Deegan, S. Chandrasekharan, and M. Angrist, "The dangers of diagnostic monopolies," *Nature*, 2009, 458(7237): 405–6.

86. https://www.counsyl.com/about/counsyl/.

87. https://www.counsyl.com/.

88. https://www.counsyl.com/campaign/end-preventable-genetic-disease/.

89. M. Hintermair and J. A. Albertini, "Ethics, deafness, and new medical technologies," *Journal of Deaf Studies and Deaf Education*, 2005, 10(2): 184–92.

90. D. Dobbs, "The science of success," *Atlantic*, December 2009.

91. https://www.counsyl.com/about/counsyl/.

92. W. Olaya, P. Esquivel, J. H. Wong et al., "Disparities in BRCA testing: When insurance coverage is not a barrier," *American Journal of Surgery*, 2009, 198(4): 562–65.

93. Email from Linda Avey, February 19, 2009.

94. http://www.aclu.org/free-speech_womens-rights/aclu-challenges-patents-breast-cancer-genes.

95. T. Ray, "Myriad files motion to dismiss ACLU's 'thinly veiled' anti-gene patenting case," *Pharmacogenomics Reporter*, August 5, 2009.

96. http://www.ascp.org/MainMenu/AboutASCP/Newsroom/NewsReleases/Gene-Patent-Class-Action-Lawsuit-Proceeds.aspx.

97. http://www.cap.org/apps/portlets/contentViewer/show.do?printFriendly=true&contentReference=cap_today%2Ffeature_stories%2F0906Gene.html.

98. E. Matloff and A. Caplan, "Direct to confusion: Lessons learned from marketing BRCA testing," *American Journal of Bioethics*, 2008, 8(6): 5–8.

99. http://www.aclu.org/free-speech-technology-and-liberty-womens-rights/association-molecular-pathology-et-al-v-uspto-et-al.

100. Interview with George Church, February 2, 2009.

101. Instant message from Jason Bobe, October 13, 2009.

102. https://ccr.coriell.org/Sections/Search/Sample_Detail.aspx?Ref=GM21677&PgId=166.

103. Amy McGuire seminar: Confidentiality of Genetic Data in Human Research, National Institute of Environmental Health Sciences, June 15, 2007.

104. H. Keyserling and T. Duerr, *South American Meditations on Hell and Heaven in the Soul of Man* (New York: Harper, 1932).

105. Email from the author to the Personal Genome Project, February 26, 2009.

CHAPTER 11: "SOMETHING MAGICAL"

1. Interview with George Church, February 2, 2009.

2. Interview with George Church, August 11, 2008.

3. Interview with George Church, December 14, 2009.

4. Interview with George Church, August 11, 2008.

5. Interview with George Church, October 20, 2008.

6. E. Pilkington, "I am creating artificial life, declares US gene pioneer," *Guardian*, October 6, 2007.

7. S. Connor, "Nobel scientist happy to 'play God' with DNA," *Independent* (U.K.), May 17, 2000.

8. Interview with George Church, February 2, 2009.

9. M. Siderits, *Buddhism as Philosophy: An Introduction* (Indianapolis: Ashgate, 2007).

10. Interview with George Church, March 16, 2009.

11. Interview with George Church, October 3, 2008.

12. http://www.bloomberg.com/apps/news?pid=conewsstory&refer=conews&tkr=ABI%3AUS&sid=aDNAix404KIA.

13. J. Karow, "Life Tech's IP-infringement suit against Illumina to go to trial in July 2011," *In Sequence*, November 17, 2009.

14. Interview with Rade Drmanac, October 9, 2007.

15. Ibid.

16. Ibid.

17. Interview with Rade Drmanac, December 8, 2008.

18. "A changing drug supply," *Nature*, 2007, 445(7127): 460.

19. J. D. McPherson, "Next-generation gap," *Nature Methods*, 2009, 6(11 Suppl): S2–5.

20. Interview with Rade Drmanac, December 8, 2008.

21. Ibid.

22. K. Davies, "Complete Genomics targets 'the first $1000 genome,'" *Bio-IT World*, November 12, 2008.

23. Interview with Rade Drmanac, December 8, 2008.

24. Ibid.

25. Interview with George Church, October 20, 2008

26. J. Karow, "Complete Genomics raises $45M in Series D round; sequencing center scheduled to open in January," *In Sequence*, August 25, 2009.

27. Ibid.

28. Interview with Rade Drmanac, December 8, 2008.

29. Interview with George Church, October 20, 2008.

30. Interview with Rade Drmanac, December 8, 2008.

31. Interview with Rade Drmanac, October 7, 2008.

32. A. Pollack, "The race to read genomes on a shoestring, relatively speaking," *New York Times*, February 9, 2008.

33. J. Karow, "PacBio to start selling next-gen sequencer to early users in 2010; goal is 100 GB/hour," *In Sequence*, February 25, 2008.

34. E. C. Hayden, "Human genomes in minutes?" Nature.com, November 20, 2008.

35. Interview with Chad Nusbaum, February 2, 2009.

36. Interview with Steve Turner, December 21, 2008.

37. Interview with Steve Turner, December 20, 2008.

38. C. H. Arnaud, "DNA sequencing forges ahead," *Chemical & Engineering News*, December 14, 2009.

39. J. Karow, "PacBio sequences E. coli genome, increases average read length to nearly 1,000 bases," *In Sequence*, February 10, 2009.

40. J. Karow, "PacBio to expand apps for SMRT platform; initial focus on cancer, infectious disease sequencing," *In Sequence*, September 22, 2009.

41. J. Eid, A. Fehr, J. Gray, et al., "Real-time DNA sequencing from single polymerase molecules," *Science*, 2009, 323(5910): 133–38.

42. "The 15-minute genome: Faster, cheaper genome sequencing on the way," *ScienceDaily*, July 29, 2009.

43. M. Herper, "The new, fast gene machine," *Forbes*, September 17, 2009.

44. Interview with Steve Turner, September 16, 2009.

45. J. Karow, "PacBio to expand apps for SMRT platform; initial focus on cancer, infectious disease sequencing," *In Sequence*, September 22, 2009.

46. J. Karow, "Survey: Illumina, SOLiD, and 454 gain ground in research labs; most users mull additional purchases," *In Sequence*, January 19, 2010.

47. Interview with George Church, October 20, 2008.

48. Cold Spring Harbor Biology of Genomes Meeting, ELSI Panel, May 8, 2008.

49. Ibid.

50. http://www.genome.gov/27026551.

51. http://www.worldsciencefestival.com/video/your-biological-biography.

52. T. Feder, "World Science Festival in NYC," *Physics Today*, 2008, 61(5): 25.

53. R. Irion, "Vera Rubin—the bright face behind the dark sides of galaxies," *Science*, 2002, 295(5557): 960–61.

54. Email from Jim Evans, May 2008.

55. Conversation with Sir Paul Nurse, May 31, 2008.

56. E. Dolgin, "The science of storytelling," *Scientist* NewsBlog, June 18, 2009.

57. Conversation with Francis Collins, May 31, 2008.

58. Interview with Francis Collins, October 11, 2008.

59. Ibid.

60. Ibid.

61. Ibid.

62. Ibid.

63. Ibid.

64. A. Sullivan, "Helping Christians reconcile God with science," *Time*, May 2, 2009.

65. F. S. Collins, *The Language of Life: DNA and the Revolution in Personalized Medicine* (New York: HarperCollins, 2009).

66. J. Kaiser, "Nominations: White House taps former genome chief Francis Collins as NIH director," *Science*, 2009, 325(5938): 250–51.

67. http://www.hhs.gov/news/press/2009pres/08/20090807d.html.

68. T. Kirby, "Francis Collins: Director of the US National Institutes of Health Profile," *Lancet*, 2009, 374(9694): 969; J. Kaiser and F. Collins, "Newsmaker interview Francis Collins: Looking beyond the funding deluge," *Science*, 2009, 326(5950): 214.

69. F. S. Collins, "The case for a US prospective cohort study of genes and environment," *Nature*, 2004, 429(6990): 475–77.

70. T. Kohnlein and T. Welte, "Alpha-1 antitrypsin deficiency: Pathogenesis, clinical presentation, diagnosis, and treatment," *American Journal of Medicine*, 2008, 121(1): 3–9.

71. S. Lueck, "One-man gridlock: Meet Tom Coburn, Senate's 'Dr. No,'" *Wall Street Journal*, December 21, 2007, http://online.wsj.com/public/article_print/SB119820693514244309.html.

72. Conversation with Dennis Pollock, July 16, 2009.

73. Interview with Francis Collins, October 11, 2008.

CHAPTER 12: CHARITY BEGINS AT HOME

1. D. B. Goldstein, *Jacob's Legacy: A Genetic View of Jewish History* (New Haven, Conn.: Yale University Press, 2008).

2. A. C. Need, D. Kasperaviciute, E. T. Cirulli, and D. B. Goldstein, "A genome-wide genetic signature of Jewish ancestry perfectly separates individuals with and without full Jewish ancestry in a large random sample of European Americans," *Genome Biology*, 2009, 10(1): R7.

3. D. B. Goldstein, "The genetics of human drug response," *Philosophical Transactions of the Royal Society of London*, 2005, 360(1460): 1571–72.

4. D. Ge, J. Fellay, A. J. Thompson, et al., "Genetic variation in IL28b predicts hepatitis c treatment-induced viral clearance," *Nature*, 2009, 461(7262): 399–401; D. B. Goldstein, "Genomics and biology come together to fight HIV," *PLoS Biology*, 2008, 6(3): e76; E. L. Heinzen, W. Yoon, S. K. Tate, et

al., "Nova2 interacts with a cis-acting polymorphism to influence the proportions of drug-responsive splice variants of SCNLa," *American Journal of Human Genetics*, 2007, 80(5): 876–83; G. L. Cavalleri, M. E. Weale, K. V. Shianna, et al., "Multicentre search for genetic susceptibility loci in sporadic epilepsy syndrome and seizure types: A case-control study," *Lancet Neurology*, 2007, 6(11): 970–80.

5. D. B. Goldstein, G. L. Cavalleri, and K. R. Ahmadi, "The genetics of common diseases: 10 million times as hard," *Cold Spring Harbor Symposia on Quantitative Biology*, 2003, 68: 395–401.

6. J. Fellay, K. V. Shianna, D. Ge, et al., "A whole-genome association study of major determinants for host control of HIV-1," *Science*, 2007, 317(5840): 944–47.

7. T. A. Manolio, "Collaborative genome-wide association studies of diverse diseases: Programs of the NHGRI's Office of Population Genomics," *Pharmacogenomics*, 2009, 10(2): 235–41.

8. N. Wade, "A dissenting voice as the genome is sifted to fight disease," *New York Times*, September 16, 2008.

9. B. Maher, "Personal genomes: The case of the missing heritability," *Nature*, 2008, 456(7218): 18–21; T. A. Manolio, F. S. Collins, N. J. Cox, et al., "Finding the missing heritability of complex diseases," *Nature*, 2009, 461(7265): 747–53.

10. https://www.23andme.com/you/journal/multiplesclerosis/overview/.

11. http://www.snpedia.com/index.php/Multiple_sclerosis.

12. J. A. Lincoln and S. D. Cook, "An overview of gene-epigenetic-environmental contributions to MScausation," *Journal of the Neurological Sciences*, 2009, 286(1–2): 54–7; L. Fugger, M. A. Friese, and J. I. Bell, "From genes to function: The next challenge to understanding multiple sclerosis," *Nature Reviews Immunology*, 2009, 9(6): 408–17.

13. D. B. Goldstein, "Common genetic variation and human traits," *New England Journal of Medicine*, 2009, 360(17): 1696–98.

14. S. P. Dickson, K. Wang, I. Krantz, H. Hakonarson, and D. B. Goldstein, "Rare variants create synthetic genome-wide associations," *PLoS Biology*, 2010, 8(1): e1000294.

15. E. T. Cirulli and D. B. Goldstein, "Uncovering the roles of rare variants in common disease through whole-genome sequencing," *Nature Reviews Genetics*, 2010, 11: 415–425.

16. Interview with Kevin Shianna, December 29, 2008.

17. Email from David Goldstein, August 14, 2009.

18. Email from Kristen Linney, September 8, 2009.

19. J. Karow, "Defective sequencing reagents to cost Illumina up to $23M in 2009; company says problem is solved," *In Sequence*, November 3, 2009.

20. Interview with Esther Dyson, September 28, 2009.

21. D. E. Jonas and H. L. McLeod, "Genetic and clinical factors relating to warfarin dosing," *Trends in Pharmacological Sciences*, 2009, 30(7): 375–86.

22. Z. E. Sauna, I. W. Kim, and S. V. Ambudkar, "Genomics and the mechanism of P-glycoprotein (ABCBL)," *Journal of Bioenergetics & Biomembranes*, 2007, 39(5–6): 481–87.

23. M. Uhr, A. Tontsch, C. Namendorf, et al., "Polymorphisms in the drug transporter gene ABCB1 predict antidepressant treatment response in depression," *Neuron*, 2008, 57(2): 203–9.

24. Q. Liu, J. Yu, Q. L. Mao-Ying, et al., "Repeated clomipramine treatment reversed the inhibition of cell proliferation in adult hippocampus induced by chronic unpredictable stress," *Pharmacogenomics Journal*, 2008, 8(6): 375–83.

25. M. L. Wong, C. Dong, J. Maestre-Mesa, and J. Licinio, "Polymorphisms in inflammation-related genes are associated with susceptibility to major depression and antidepressant response," *Molecular Psychiatry*, 2008, 13(8): 800–12.

26. G. Laje, S. Paddock, H. Manji, et al., "Genetic markers of suicidal ideation emerging during citalopram treatment of major depression," *American Journal of Psychiatry*, 2007, 164(10): 1530–38.

27. http://www.informatics.jax.org/searches/accession_report.cgi?id=MGI: 88145.

28. R. H. Lipsky and A. M. Marini, "Brain-derived neurotrophic factor in neuronal survival and behavior-related plasticity," *Annals of the New York Academy of Sciences*, 2007, 1122: 130–43; P. Bekinschtein, M. Cammarota, I. Izquierdo, and J. H. Medina, "BDNF and memory formation and storage," *Neuroscientist*, 2008, 14(2): 147–56.

29. J. Fan and P. Sklar, "Genetics of bipolar disorder: Focus on BDNF Val66Met polymorphism," *Novartis Foundation Symposium*, 2008, 289: 60–72; discussion 72–73, 87–93.

30. J. O. Groves, "Is it time to reassess the BDNF hypothesis of depression?" *Molecular Psychiatry*, 2007, 12(12): 1079–88.

31. J. K. Rybakowski, "BDNF gene: Functional Val66Met polymorphism in mood disorders and schizophrenia," *Pharmacogenomics*, 2008, 9(11): 1589–93.

32. A. R. Brunoni, M. Lopes, and F. Fregni, "A systematic review and meta-analysis of clinical studies on major depression and BDNF levels: Implications

for the role of neuroplasticity in depression," *International Journal of Neuro-psychopharmacology*, 2008, 11(8): 1169–80; H. Y. Lee and Y. K. Kim, "Plasma brain-derived neurotrophic factor as a peripheral marker for the action mechanism of antidepressants," *Neuropsychobiology*, 2008, 57(4): 194–99.

33. Emails from David Goldstein and Anna Need, January 3, 2010.

34. Interview with Kevin Shianna, December 29, 2008, October 13, 2009.

35. Ibid.

36. Ibid.

37. Ibid.

38. Ibid.

39. Interview with Jason Smith, October 21, 2009.

40. Ibid.

41. Ibid.

42. Ibid.

43. Interview with Kevin Shianna, December 29, 2008, October 13, 2009.

44. http://people.genome.duke.edu/~dg48/index.php

45. D. Ge, K. Zhang, A. C. Need et al., "WGAViewer: Software for genomic annotation of whole genome association studies," *Genome Research*, 2008, 18(4): 640–43.

46. J. Karow, "Duke to sequence 50 human genomes on Illumina GA; plans more large sequencing studies," *In Sequence*, January 20, 2009.

47. Interviews with Dongliang Ge, October 2009.

48. Ibid.

49. Ibid.

50. Ibid.

51. Ibid.

52. J. J. Michiels, A. Gadisseur, U. Budde, et al., "Characterization, classification, and treatment of von Willebrand diseases: A critical appraisal of the literature and personal experiences," *Seminars in Thrombosis & Hemostasis*, 2005, 31(5): 577–601.

53. Interviews with Dongliang Ge, October 2009.

54. Ibid.

55. Ibid.

56. Interview with Hugh Rienhoff, December 7, 2008.

57. http://www.snpedia.com/index.php/SNPedia:FAQ#How_many_SNPs_are_in_SNPedia.3F.

58. http://www.ncbi.nlm.nih.gov/projects/SNP/snp_summary.cgi.

59. Interview with Greg Lennon, December 17, 2008.

60. http://wiki.github.com/xwu/trait-o-matic.

61. http://www.pharmgkb.org/.

62. Email from Sasha Zaranek, December 1, 2009.

63. P. D. Stenson, M. Mort, E. V. Ball, et al., "The Human Gene Mutation Database: 2008 update," *Genome Medicine*, 2009, 1(1): 13.

64. Email from Mike Cariaso, October 22, 2009.

65. http://www.ncbi.nlm.nih.gov/omim/

66. http://www.genetests.org.

67. M. Angrist, "We are the genes we've been waiting for: Rational responses to the gathering storm of personal genomics," *American Journal of Bioethics*, 2009, 9(6): 30–31.

68. A. Barton and J. Worthington, "Genetic susceptibility to rheumatoid arthritis: An emerging picture," *Arthritis & Rheumatism*, 2009, 61(10): 1441–46.

69. T. A. Cooper, L. Wan, and G. Dreyfuss, "RNA and disease," *Cell,* 2009, 136(4): 777–93; G. S. Wang and T. A. Cooper, "Splicing in disease: Disruption of the splicing code and the decoding machinery," *Nature Reviews Genetics*, 2007, 8(10): 749–61.

70. R. Alcalai, J. G. Seidman, and C. E. Seidman, "Genetic basis of hypertrophic cardiomyopathy: From bench to the clinics," *Journal of Cardiovascular Electrophysiology*, 2008, 19(1): 104–10.

71. Email from Steven Pinker, September 29, 2009.

CHAPTER 13: ANTARCTICA

1. Email from Abraham Rosenbaum, August 25, 2009.

2. Interview with Abraham Rosenbaum, June 11, 2009.

3. http://wiki.github.com/xwu/trait-o-matic.

4. K. Gronskov, J. Ek, and K. Brondum-Nielsen, "Oculocutaneous albinism," *Orphanet Journal of Rare Diseases,* 2007, 2: 43; T. Suzuki and Y. Tomita, "Recent advances in genetic analyses of oculocutaneous albinism types 2 and 4," *Journal of Dermatological Science*, 2008, 51(1): 1–9.

5. Interview with Hugh Rienhoff, May 20, 2008.

6. L. Stanchina, V. Baral, F. Robert, et al., "Interactions between Sox10, Edn3 and Ednrb during enteric nervous system and melanocyte development," *Developmental Biology*, 2006, 295(1): 232–49; N. M. Le Douarin, S. Creuzet, G. Couly, and E. Dupin, "Neural crest cell plasticity and its limits," *Development*, 2004, 131(19): 4637–50.

7. G. L. Moldovan and A. D. D'Andrea, "How the Fanconi anemia pathway guards the genome," *Annual Review of Genetics*, 2009, 43: 223–49.

8. http://research.nhgri.nih.gov/projects/bic/Member/cgi-bin/bic_full_sum mary.cgi?table=brca1_exons.

9. http://research.nhgri.nih.gov/projects/bic/Member/cgi-bin/bic_full_sum mary.cgi?table=brca2_exons.

10. R. C. Panguluri, L. C. Brody, R. Modali, et al., "BRCA1 mutations in African Americans," *Human Genetics*, 1999, 105(1–2): 28–31.

11. Email from Mary-Claire King, October 28, 2009.

12. B. Yngvadottir, Y. Xue, S. Searle, et al., "A genome-wide survey of the prevalence and evolutionary forces acting on human nonsense SNPs," *American Journal of Human Genetics*, 2009, 84(2): 224–34.

13. Interview with Andrea Loehr, June 8, 2009.

14. Interview with Hugh Rienhoff, December 9, 2008.

15. K. Sanneh, "Discriminating tastes," *New Yorker*, August 10, 2009.

16. http://www.fas.harvard.edu/~amciv/faculty/gates.shtml.

17. http://www.pbs.org/wnet/facesofamerica/.

18. Interview with Henry Louis Gates, Jr., January 7, 2010.

19. Interviews with Henry Louis Gates, Jr., January 7, 2010, and Rick Kittles, January 30, 2010.

20. Interview with Henry Louis Gates, Jr., January 7, 2010.

21. Ibid.

22. Interview with Rick Kittles, January 30, 2010.

23. Interview with Henry Louis Gates, Jr., January 7, 2010.

24. Ibid.

25. Interview with George Church, November 24, 2009.

26. J. H. Relethford, "Genetic evidence and the modern human origins debate," *Heredity*, 2008, 100(6): 555–63.

27. M. Via, E. Ziv, and E. G. Burchard, "Recent advances of genetic ancestry testing in biomedical research and direct to consumer testing," *Clinical Genetics,* 2009, 76(3): 225–35.

28. M. D. Shriver and R. A. Kittles, "Genetic ancestry and the search for personalized genetic histories," *Nature Reviews Genetics*, 2004, 5(8): 611–18.

29. http://spittoon.23andme.com/2009/11/19/introducing-relative-finder-the-newest-feature-from-23andme/.

30. Interview with Henry Louis Gates, Jr., January 7, 2010.

31. E. Lott, "Criticism in the vineyard: Twenty years after 'Race,' Writing, and Difference," *PMLA*, 2008, 123(5): 1522–27.

32. D. A. Bolnick, D. Fullwiley, T. Duster, et al., "Genetics. The science and business of genetic ancestry testing," *Science,* 2007, 318(5849): 399–400; A. Nordgren and E. Juengst, "Can genomics tell me who I am? Essentialistic rhetoric in direct-to-consumer DNA testing," *New Genetics & Society,* 2009, 28(2): 157–72.

33. Interview with Henry Louis Gates, Jr., January 7, 2010.

34. T. Ray, "Illumina delivers personal genome sequencing results to first consumer: Amadeus capital's Hauser," *Pharmacogenomics Reporter,* September 2, 2009.

35. Interview with Henry Louis Gates, Jr., January 7, 2010.

36. Interview with Rick Kittles, January 30, 2010.

37. Interview with Duana Fullwiley, March 10, 2008.

38. Interview with Henry Louis Gates, Jr., January 7, 2010.

39. Ibid.

40. N. Scola, "Crowdsourcing the genome," *SEED,* February 27, 2009.

41. M. P. O'Donnell, "The genetic information nondiscrimination act—a wake-up call: Great intentions, but a setback for health impact and cost-effectiveness of workplace health promotion," *American Journal of Health Promotion,* 2010, 24(3): iv–v.

42. M. Angrist, "Eyes wide open: The personal genome project, citizen science and veracity in informed consent," *Personalized Medicine,* 2009, 6(6): 691–99.

43. http://investor.illumina.com/phoenix.zhtml?c=121127&p=IROL-newsArticle&ID=1298128.

44. http://venturebeat.com/2007/11/29/got-350000-knome-has-your-full-genome/.

45. C. Ayres, "Personal genome sequence is must-have accessory for billionaires, says Knome head," *Times of London,* November 14, 2009.

46. R. Drmanac, A. B. Sparks, M. J. Callow, et al., "Human genome sequencing using unchained base reads on self-assembling DNA nanoarrays," *Science,* 2009.

47. "Complete Genomics nears commercial launch; eyes thousands of genomes in 2010," *GenomeWeb Daily News,* January 14, 2010.

48. http://investor.illumina.com/phoenix.zhtml?c=121127&p=IROL-newsArticle&ID=1374339&highlight=.

49. K. Davies, "It's 'Watson meets Moore' as ion torrent introduces semiconductor sequencing," *Bio-IT World,* March 1, 2010.

50. arep.med.harvard.edu/gmc/ppt/09Jun9_CGC_Hynes.pdf.

51. Interview with Maynard Olson, May 2, 2008.

52. A. Kong, V. Steinthorsdottir, G. Masson, et al., "Parental origin of sequence variants associated with complex diseases," *Nature*, 2009, 462(7275): 868–74.

53. http://www.youtube.com/watch?v=8qxMbP2gWpU.

EPILOGUE: HERE IS A HUMAN BEING

1. http://www.technologyreview.com/blog/editors/25125.

2. Email from Jason Bobe, January 5, 2010.

3. Interview with George Church, June 9, 2009.

4. Email from Jason Bobe, August 12, 2009.

5. M. J. Khoury, C. M. McBride, S. D. Schully, et al., "The Scientific Foundation for personal genomics: Recommendations from a National Institutes of Health–Centers for Disease Control and Prevention multidisciplinary workshop," *Genetics in Medicine*, 2009, 11(8): 559–67.

6. http://kara.allthingsd.com/20090904/23andme-co-founder-linda-avey-leaves-start-up-to-focus-on-alzheimers-research/.

7. Interview with Dietrich Stephan, June 10, 2009.

8. Interview with Linda Avey, November 17, 2009.

9. http://www.ignitehealth.org/about/leadership.asp.

10. J. Karow, "Ignite Institute to install 100 ABIO SOLiD 4 systems by year end under life tech collaboration," *In Sequence*, January 28, 2010.

11. Interview with Dietrich Stephan, June 10, 2009.

12. J. Kaiser, "Biotechnology. Bankruptcy won't stop decode, says its founder, Stefansson," *Science*, 2009, 326(5957): 1172.

13. M. Carmichael, "The world's most successful failure," *Newsweek*, February 12, 2010.

14. http://www.techcrunch.com/2009/10/29/layoffs-confirmed-at-23andme/.

15. B. Japsen and S. M. Jones, "Walgreens postpones carrying Pathway Genomics genetic test kit," *Los Angeles Times*, May 13, 2010.

16. http://energycommerce.house.gov/index.php?option=com_content&view=article&id+2009:committee-investigates-personal-genetic-testing-kits&catid=12:media-advisories&Itemid=55.

17. http://www.techcrunch.com/2009/12/23/23andme-funding/.

18. Email from George Church, September 7, 2009.

19. G. M. Church, "Genomes for all," *Scientific American*, 2006, 294(1): 46–54.

Index